National Audubon Society
Field Guide to
North American
Weather

A Chanticleer Press Edition

National Audubon Society
Field Guide to
North American
Weather

David M. Ludlum
Founder of *Weatherwise* magazine

Alfred A. Knopf, New York

This is a Borzoi Book
Published by Alfred A. Knopf, Inc.

Copyright © 1991, 1997 by Chanticleer
Press, Inc. All rights reserved under
International and Pan-American
Copyright Conventions. Published in
the United States by Alfred A. Knopf,
Inc., New York, and simultaneously in
Canada by Random House of Canada
Limited, Toronto. Distributed by
Random House, Inc., New York.

Prepared and produced by
Chanticleer Press, Inc., New York.

Type set in Garamond by Graphic Arts
Composition, Philadelphia. Printed and
bound by Dai Nippon, Japan.

Published September 1991
Seventh Printing, October 1997

Library of Congress Cataloging-in-
Publication Number: 91-52707
ISBN: 0-679-40851-7

CONTENTS

THE AUTHOR AND CONTRIBUTORS

Author Dr. David M. Ludlum, a renowned weather watcher and weather historian, founded *Weatherwise* magazine in 1948 and served as its editor-in-chief for thirty years. In addition, he operated a consulting business specializing in weather instruments from the end of World War II until 1983. He has written twelve books on the subject of meteorology, among them *The American Weather Book* and *The Weather Factor*. Dr. Ludlum's other works are devoted to the history of hurricanes, tornadoes, and winter weather in the United States. He continues to write about the weather and to travel around the world observing it.

Dr. Richard A. Keen reviewed the essays in this book for factual accuracy, contributed photos, helped with the selection of photographs, and wrote the text accounts for lightning, tornadoes and other whirls, snowstorms and ice storms, flood, drought, and optical phenomena. A freelance science writer, nature photographer, and public speaker, he has published books, articles, and photographs on the topics of meteorology and astronomy. He recently completed *Skywatch: The Eastern Weather Guide,* and is currently working on a book about the weather in Minnesota, and another on the weather in Alaska.

Ronald L. Holle acted as a consultant to this guide and assisted in both the selection of photographs included in the color plates and in the writing of several text accounts. He is a research meteorologist with the National Severe Storms Laboratory of the National Oceanic and Atmospheric Agency (NOAA), in Boulder, Colorado. His research has focused on meteorological studies of cloud-to-ground lightning, mesoscale cloud organization and interaction, and photographic analysis of convective clouds. He is currently president of the Denver-Boulder chapter of the American Meteorological Society, and has written many technical papers, articles, and reports on weather.

ACKNOWLEDGMENTS

Many people contributed to the preparation of this book. I am grateful to the multitude of government and private weather observers who note the vagaries of the weather; to those who devotedly compiled climatic data from the millions of weather observations performed annually; and to the tireless librarians who provided me with invaluable assistance in my research. Rendering more immediate guidance in this project were my editors at Chanticleer Press: Ann Whitman, Jane Mintzer Hoffman, and Carol Healy, whose collective editorial acumen has made a presentable volume of my raw manuscript. Photo editor Tim Allan did yeoman's work in gathering and selecting all the photographs for this guide. I also extend my gratitude to Dr. Richard A. Keen, who read the manuscript and made many helpful suggestions; and to Ronald Holle, who lent his knowledge of cloud physics to the text accounts of the cloud photographs. Other members of the Chanticleer staff helped with every stage of this book. My thanks to Barbara Balch for her design; to Gretchen Bailey Wohlgemuth, who diligently saw the book through production; to Micaela Porta and Kate Jacobs, who lent invaluable editorial support; and to Ferris Cook, who gathered much of the research material for the illustrations that accompany the essays. I would like also to extend my thanks to John Farrand, Jr., who as unofficial natural science consultant for the Chanticleer team has offered much guidance and support.

The "Dark Spot Stage of the Waterspout Life Cycle" illustration in the essay on tornadoes is based on research done by Joseph H. Golden at the National Oceanic and Atmospheric Agency (NOAA) and was developed with his assistance. The map of tornado-prone areas in the United States and Canada incorporates information provided by Michael J. Newark of the Canadian Atmospheric Environment Service. T. Theodore Fujita, of the University of Chicago, granted us permission to base our diagrams of wet and dry microbursts on his published illustrations.

I extend many thanks as well to the photographers who contributed work of such high quality to this project. Their images are instrumental not only in educating us about the weather, but also in fostering an appreciation of its beauty and power as well.

INTRODUCTION

Weather is the single most powerful and pervasive force on our planet. Rain, snow, heat, cold, drought, and storms all play a dramatic part in shaping life on earth.

Every life form, including our own, reacts to great and small changes in the weather. As surely as birds fly south in winter, we adapt our behavior (such as canceling the picnic) or our mode of life (learning to conserve water) in response to the weather's dictates. We are at its mercy, and we shall thrive or perish accordingly.

Observing Weather: Unlike birds, mammals, and flowers, the weather is always with us, in one form or another. There is no need to travel to the seashore or the forest, the countryside or a national park in search of the weather; to see it, feel it, and experience it, you need only step outdoors or look out a window any day of the year.

On the other hand, weather is much more mysterious than many other aspects of nature; and weather watching often seems to be as much an art as a science. Much of what constitutes our weather is invisible, capable of being identified or anticipated only by those in the know—or those in possession of sophisticated equipment. People who are weather-wise function as interpreters for those of us who cannot read the signs.

To add to the mystery, weather is constantly changing. Generally speaking, a bird-watcher who *can* recognize a Canada goose *will* recognize a Canada goose. A weather watcher, on the other hand, is observing a subject in a constant state of flux; predicting the weather is like learning to predict the behavior of an individual Canada goose: Will this bird land on the pond or in the cornfield? Will that thunderhead produce an electrical storm, or will it blow over?

What Is Weather?:

Briefly stated, the weather reflects the prevailing conditions in the air masses that overlie any given spot on the earth. Changes in weather are brought about by movement in those air masses, which in turn brings about changes in such important factors as *temperature*, *wind speed*, *wind direction*, *humidity*, and *atmospheric pressure*.

While the earth spins on its axis, the warmth of the sun heats the atmosphere; the area at the equator receives more heat from solar radiation than any other area, so the air there is warmer. The warmer air rises and flows toward the poles, where the atmosphere is cooler and more stable. The arrival of that warmer air at the poles sets the cooler air in motion. Thus the entire atmosphere is in constant movement. As this air flows poleward, it is further governed by an important physical force known as the *Coriolis force*, sometimes called the *Coriolis effect*. This force is a result of the earth's rotation on its axis; as air moves toward the poles, it is deflected by the spinning of the planet. Thus, instead of flowing north or south in a straight line, the air moves in a curved path—toward the east. (You can visualize this effect easily with a simple experiment. Place a sheet of paper on top of a spinning turntable and draw a line directly from the edge of the paper to the spindle at the center, or from the spindle out to the rim. Although your pen appears to move straight across the paper, the line your pen makes will invariably be curved.)

Why Learn About Weather?: Even a professional meteorologist cannot say with total accuracy what the weather will be tomorrow. The many ingredients that make up our weather are volatile, and changes can be quite dramatic.

It is possible, however, to be fairly close to the mark. You may not be able to say for certain that it will rain at four o'clock; but you will be able to state with conviction that "It looks like rain." In many cases, you will be right. For example, the approach of a low-pressure system from the tropics may mean that a tropical storm—perhaps even a hurricane—is on its way. If you see a halo around the moon at night, there is a good chance that the next day will bring snow. There is no magic in interpreting such conditions. Once you know that the moon's halo, for example, results from the refraction of light through tiny ice crystals in the upper atmosphere, then you are on firm ground in deducing that the clouds above are cold enough to generate snow.

What is more, when you learn about weather you will be more attuned to other natural events as well. If, for example, you are interested in seeing warblers during their spring migration, the weather is important. Sunlight warming the boughs of a tall tree will stir the insect life in the branches; and the activity of the insects will attract the warblers. A spell of unseasonably cool weather, however, may mean that both the spring and the warblers will be late.

In a sense, then, this field guide differs

from ordinary field guides. It will teach you not only to observe, but also to understand and interpret what you see, and to reach conclusions based on those interpretations.

Coverage and Geographical Scope: This book includes not only the major kinds of cloud formations, precipitation, and storm systems, but also such "quiet" weather events as rainbows and other optical phenomena; and weather-related crises such as floods, droughts, and forest fires. It covers all of the chief weather events and patterns common on the North American continent north of Mexico.

Organization of the Guide: The guide is divided into three main sections: essays, color plates, and text accounts.

Essays: In the first section are essays that describe, in a general way, the scientific forces and principles behind different kinds of weather. The essays are not intended to describe weather phenomena but to provide you with a simple, solid introduction to the science of meteorology, as well as a ready reference for brushing up on these principles. Fifty or so black-and-white illustrations are included to help you visualize the events described.

Color Plates: At the heart of this guide are the color photographs. They are organized according to the kind of phenomenon or event depicted: the atmosphere, seasons, cloud kinds, storms, optical phenomena, floods, and drought. The photographs show you a kind of weather that existed at one moment in time; but just as no two snowflakes are identical, so, too, are there no identical occurrences in the behavior of the atmosphere around us. The photographs present you with dramatic visual representations of events that have actually taken place, rather than stylized conceptions of the forces at work. It is impossible to say, for example, that a thunderhead will *always* look like an anvil; much depends on your point of view, the action of the wind, the age of the cloud, and the time of day and the time of year. On the other hand, if you study the photographs in conjunction with the text and your own experience, you will quickly learn how nature echoes and recreates the scenes depicted. The caption under each photograph gives the plate number, identifies the weather phenomenon pictured, and tells you the page number of the relevant text account.

At the close of the color section is a selection of black-and-white images of memorable, historically important, or meteorological events.

Text Accounts: The second main text section consists of vignettes of the meteorological events depicted in the photographs. Each account in this section is keyed to the color plates, which immediately precede this portion of the text. Some photographs depict more than one kind of event or phenomenon—for example, a photograph may show two different kinds of clouds. Because weather is not precise but always changing, many text accounts are illustrated by more than one photograph.

Each text account starts off with a *description* of the event or phenomenon seen in the photo. There follows next a brief overview of the *environment*—the conditions that, generally speaking, will give rise to that kind of event. The *season* and *range* are noted, and these sections are followed by one on *variations*—assorted possibilities for the shape, color, duration, level of organization, and so forth, of the object or event depicted. The paragraph entitled *significance* gives, where appropriate, the possible or likely results or aftereffects of the object or event in the photograph; in some cases, the relevance to human activity and human safety is also described. The *comments* paragraph, when present, provides anecdotes and general information about significant storms, historical events, and the like.

Using the Thumb Tab Guide: The organization of the color plates is outlined in a table called the *Key to the Color Plates.*

Each group of color plates is designated by a distinctive thumb tab—a graphic symbol representing the kind of events depicted within that group. In the Key to the Color Plates, each thumb-tab symbol (at left) is listed in order of its occurrence in the book. On the right is a summary of the kind of event or weather represented by the symbol and a list of color plates included in that group. Within the color-plate section, each group is set off by the use of that symbol in the thumb tab on the left-hand page.

Our purpose in preparing this book is to provide a deeper and more pleasurable understanding of the forces of nature that shape our activities and determine our survival, to help you learn to read and interpret the signs of the weather around you—some of the basic language of nature.

Peruse the introductory essays in the front of the book, study the color photographs, and read the individual text accounts of the events depicted in each photograph. Then take a walk outside—in the sunshine or, perhaps, even in the rain.

HOW TO USE THIS GUIDE

Example 1
Gray to pale blue water-droplet clouds at middle levels

While camping in southeastern Utah, near the Colorado border, you notice that a wind has blown in from the southwest, covering the sky with a dense, unvarying layer of pale blue clouds. The sun shines dimly from behind the thick layer, which appears to be very smooth and lacks any kind of wisps or streamers.

1. Turn to the Thumb Tab Guide immediately preceding the color plates and look for the symbol that most closely resembles the kind of cloud you are looking at. In the group called Middle Clouds you find a silhouette that seems very similar; the key refers you to color plates 35–54.

2. Turn to the color plates. A quick comparison shows that the clouds you are looking at most closely resemble the photographs of altostratus clouds. The captions refer you to the text accounts for altostratus clouds, in the section Middle Clouds, beginning on page 452.

3. You read the text and discover that altostratus clouds indicate moisture at middle levels in the atmosphere (above 6,500 feet/2,000 m). They may precede a storm; and in the Southwest, in summer, they indicate the possibility of heavy rain or perhaps a thunderstorm in the near future.

Example 2
A small funnel forming at the base of a cloud

You are visiting friends in Kansas who live just a few miles from your house, and you have heard on the radio that a tornado watch has been issued. You scan the sky and notice that a small funnel has formed at the base of a thick thunderstorm cloud overhead.

1. Turn to the Thumb Tab Guide. There you find a vortex-shaped silhouette, standing for the group Tornadoes and Other Whirls. The symbol refers you to the color plates 207–230.

2. You look at the color plates and quickly surmise that the funnel shape in the sky overhead may be the beginning stage of a tornado. The captions refer you to the text on pages 511–522.

3. Reading the text, you become convinced that there is a tornado forming. You and your friends immediately seek the storm shelter and wait for the threat to pass.

Example 3
A bright ring of pale, colorful light circling the moon

You are walking down a country lane on a winter's night in the Northeast. Overhead is a full moon, its brightness somewhat lessened by a thin, wispy layer of clouds. Around the moon is a ring of pale red and blue.

1. Scanning the Thumb Tab Guide, you find the symbol for Optical Phenomena; within that group you see a silhouette showing a ring around the sun or moon. You turn to the color plates 335, 336, 338, and 339.

2. A quick comparison shows that you are looking at a halo around the moon. The captions refer you to the text on pages 548–562.

3. The text explains that haloes are caused by the refraction of light through ice crystals present in the cloud layer, most often in cirriform clouds, at temperatures well below freezing. Under the heading Significance, you read that in your area at this time of year, the presence of a halo means that a snowstorm may arrive within the next day or two.

Part I
Essays

THE ATMOSPHERE

The atmosphere is our most valuable possession. Without its thin, protective envelope of gases and foreign particles held in delicate balance, the earth could not sustain human, animal, or vegetable life. The earth's current, mostly benign, climate results from a number of factors working in our behalf, among them the atmosphere's general patterns of circulation, thermal structure, distribution of moisture, sequence of storms and fair-weather periods, and its reaction to the introduction of foreign particles. Only with a more complete understanding of the nature and behavior of our atmosphere can we make intelligent decisions in the present age and preserve it for future generations.

The presence of a sizable amount of oxygen as a major component of the atmosphere enables the earth to sustain life. Another distinguishing component of our air is the occurrence of varying amounts of water vapor. Rising moisture in the atmosphere gives birth to many forms of meteorological action: clouds, fog, rain showers, cyclonic storms, thunderstorms, hurricanes, tornadoes, snowstorms, blizzards, and floods. These weather phenomena will be discussed in the essays of this guide.

Structure of the Atmosphere

| | miles |
| | km |

temperature ·········· - - - - - - - - - - - - - 68.2
                                                  110

aurora
thermosphere - - - - - - - - - - - - - - - - - - 62
                                                  100

- 56
                                                  90

mesopause - 49.6
                                                  80

- 43.4
                                                  70

mesosphere

- 37.2
                                                  60

- 31
                                                  50

stratopause

- 24.8
                                                  40

- 18.6
                                                  30

stratosphere

ozone - 12.4
                                                  20

- 6.2
tropopause   10

troposphere

| -148 | -112 | -76 | -40 | -4 | 32 | 68 | °F |
| -100 | -80 | -60 | 40 | -20 | 0 | 20 | °C |

Depth: The gaseous envelope called the atmosphere is relatively thin in comparison with the diameter of the earth, yet it is thick enough to protect us from deadly rays and from fiery objects originating in outer space. Its chief characteristic is that it decreases in density, or mass per unit volume, with increasing altitude. As a result, the pressure exerted on the earth by air particles also decreases with altitude. At any level in the atmosphere, the pressure is essentially the weight of the overlying air, so as one rises above more of the air, the pressure decreases. Thus, most of the atmosphere's mass lies in its very lowest levels—about 99 percent within 18 miles of the earth's surface, and 99.999 percent in the lowest 50 miles. Even 50 miles up, the atmosphere is still dense enough to intercept incoming solar rays and scatter them in all directions, but there is essentially no meteorological activity. So for practical purposes, the atmosphere can be considered to be no more than about 50 miles thick.

Structure: A knowledge of the atmosphere's structure is essential to an understanding of daily weather events. (See the drawing of the structure of the atmosphere on page 24.) The lower portion, called the troposphere, is the region where vertical mixing of hot and cold currents takes place, often attended by much turbulence. In this zone, temperature usually decreases with increasing altitude. The altitude of the troposphere's upper limit, called the tropopause, varies with season and latitude. It is highest where the air is warmest—namely, in the tropics and equatorial zone—and during the summer months elsewhere. On the average, the tropopause lies about 12 miles above the equator and only 5 miles above the poles. These figures vary a few miles either way with the movement of warm and cold airstreams around the globe.

The tropopause marks the meeting place of the troposphere and the stratosphere. It is the level at which air temperature ceases to drop steadily with altitude and either remains constant or starts increasing slightly. Rising warm air currents will continue moving upward in the troposphere because they are surrounded by cooler air. But when they reach an environment where the temperature remains steady or increases, the dynamic upward push of the surrounding air ceases to operate, and the current rises no more. The tropopause, therefore, marks the "ceiling" of the lower atmosphere and confines almost all significant weather to the zone of the troposphere.

Chemical Composition: Under ordinary circumstances, the basic composition of the troposphere and stratosphere does not change—nitrogen makes up about 78 percent and oxygen about 21 percent by volume. The remaining 1 percent includes trace gases like argon, helium, and several others whose concentrations vary from time to time and from place to place. Gases such as water vapor, carbon dioxide, carbon monoxide, sulfur dioxide, nitrogen dioxide, methane, and ozone, although present in only small amounts, are important in sustaining life. In addition, varying amounts of foreign particles, composed of dust, salt, pollen, chemicals, volcanic ejections, and meteoric fragments, are occasionally found in the atmosphere.

The layer directly above the tropopause is the stratosphere. Here the temperature remains nearly constant through the first 18 miles (29 km), then begins a rather sharp increase that continues up to the stratopause, about 30 miles (48 km) above the earth's surface at the top of the stratosphere. High temperatures occur in the stratosphere because of the presence of atmospheric ozone, confined to heights between 6 and 30 miles (10 and 50 km). Ozone grows hotter when it absorbs ultraviolet radiation from the sun.

Pollution and Protection: Some industrial chemicals, such as chlorofluorocarbons, as well as volcanic dust, can find their way into the stratosphere. Both pollutants may attack our protective ozone shield, which absorbs ultraviolet radiation in the upper atmosphere, thus reducing the amount that reaches the surface of the earth. Exposures to increased doses of ultraviolet rays would cause severe sunburn and increase the risk of skin cancers.

Above the stratosphere, directly over the stratopause, lies a zone called the mesosphere, where temperatures again decrease with altitude to a level of about 50 miles (80 km) above the earth. A thin boundary zone called the mesopause separates the mesosphere from the next atmospheric level, the thermosphere, where the temperature reverses once again, increasing with height up to about 120 miles (193 km). Air density in the thermosphere is extremely low, and very little energy from solar radiation is needed to produce the observed temperature increases.

Still another zone of the atmosphere, called the ionosphere, extends upward to about 250 miles (402 km). Here the molecules and atoms of air become ionized, or electrically charged, by absorption of high-energy radiation from the sun. The chief role of the ionosphere as it pertains to our daily lives is to reflect radio transmissions back to earth, permitting the reception of long-distance radio broadcasts. No meteorological activity takes place in the ionosphere, but it is responsible for

Radiation and Rotation of the Earth

Coriolis Force

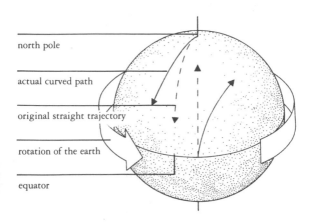

north pole

actual curved path

original straight trajectory

rotation of the earth

equator

Solar Radiation

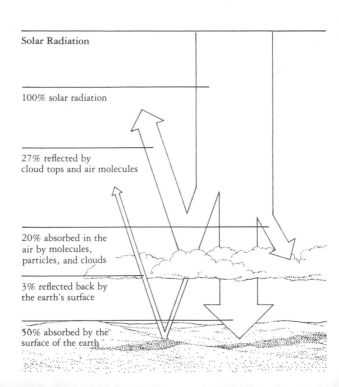

100% solar radiation

27% reflected by
cloud tops and air molecules

20% absorbed in the
air by molecules,
particles, and clouds

3% reflected back by
the earth's surface

50% absorbed by the
surface of the earth

the phenomena attending the aurora borealis, or northern lights. These intriguing displays are caused by high-energy particles sent to earth from flares on the sun. Such solar particles further ionize the oxygen and nitrogen gases of the ionosphere, causing them to emit light in the form of a glow over the polar regions.

Heat: In the atmosphere, heat is transferred by one of three processes: conduction, convection, or radiation.

Conduction is the movement of heat by and through a substance by stimulation of its molecular activity without any movement of the matter itself. The ability of substances to conduct heat varies considerably. For example, the handle of a solid metal spoon heats rapidly when the other end is immersed in a hot liquid, whereas the handle of a wooden spoon remains cool. This difference comes about because wood is a relatively poor conductor. The atmosphere is also a poor conductor, so conduction plays a minor role in the warming or cooling of air.

Of much greater importance in weather events is convection, or the transport of heat by the movement of parcels of air. Convection is usually described as taking place vertically. Fair-weather cumulus clouds, for example, are caused by rising air currents. Air heated at the surface of the earth tends to rise as an invisible column through surrounding cooler air. The rising column continues to cool in its ascent until it reaches its dew point, or saturation temperature. At this point, also called the condensation level, water vapor in the rising air becomes liquid and is visible as a cloud.

Most atmospheric heat originates on the sun and arrives by radiation, or the invisible transfer of heat by rays of various wavelengths. The conducting medium is unaffected by the passage of radiative heat through it, but the terminal object of the rays absorbs, scatters, or reflects the heat. Of the solar radiation arriving at the top of the atmosphere, only about 50 percent is absorbed by the surface of the earth. About 20 percent is absorbed in the air by molecules, particles, and clouds; 27 percent of the total is reflected back into outer space by cloud tops and air molecules. The remaining 3 percent is reflected directly back to space by the earth's surface.

Although the earth as a whole absorbs about as much heat as it loses, the rates of absorption and loss differ according to latitude. In the tropics and equatorial regions, where the daytime sun passes nearly overhead, more heat is received than is lost. The opposite is true for more poleward latitudes, where sunlight strikes the ground at lower angles. This imbalance of heat in various parts of the globe drives the winds and ocean currents, which act as giant heat pumps, transferring warm air or water from the lower latitudes poleward, and returning cold air and water to the tropics.

Moisture: Water vapor constitutes a very significant, although small, portion of the atmosphere, varying from very little in desert and polar regions to as much as 4 percent by volume in jungle tracts. Water is a major factor in the making of weather. It can exist in a gaseous, liquid, or solid state. In its gaseous state, water is an invisible vapor; in its liquid state, rain. If temperatures are below freezing, water changes directly into solid ice crystals—sleet, hail, or snow. This solid form may be converted to a liquid state by melting, or to a gaseous state by evaporation.

When water changes from one state to another, the rather mysterious exchange of latent, or "hidden," heat is involved. Unlike "sensible" heat—heat that we can feel, or sense, and measure as temperature—latent heat is stored in the molecular structure of liquid and solid water. When water vapor is cooled and condenses into a cloud or rain droplet, latent heat is released to the air, where it becomes sensible heat, raising the air's temperature. In the opposite process, when water droplets evaporate, sensible heat is removed from the air, lowering the air's temperature and becoming stored as latent heat in the vapor. This heat supplies further energy to the atmosphere.

The chief function of water vapor in the atmosphere is to supply the ingredients for the various forms of rain and snow. Obviously, water vapor can exist without being transformed into precipitation—and the precise cause of this transformation has long been a mystery to meteorologists and the object of much research during recent decades. It is now thought that a small impurity, called a condensation nucleus, is necessary to initiate growth of a raindrop or ice crystal. The nuclei come from many sources, such as blowing soil, volcanoes, smokestacks, and the spray of oceans. Some of these particles are hygroscopic; that is, they readily attract water.

Precipitation: Two processes have been identified in the formation of precipitation. The first, coalescence, takes place when tiny droplets of condensed water vapor within a cloud merge to form larger raindrops. In the turbulence within a cloud, the raindrops collide and grow larger until their weight can no longer be sustained and they fall from the cloud. The second process takes place at higher elevations, where below-freezing temperatures prevail and supercooled (below-freezing) liquid water droplets as well as ice crystals are present. Because of pressure differences between ice and water, the supercooled water is attracted to and freezes on the frozen crystals. The crystals grow into snowflakes, and when they become heavy enough they descend from the cloud. Depending on the surrounding temperatures, they may either continue to the ground as snowflakes or melt and become raindrops.

Stability and Instability of Air: When a parcel of air moves upward through the troposphere, it passes through levels of successively lower air pressures and temperatures. As a result, the unsaturated air expands and cools internally at a rate of 5.4°F for every 1,000 feet of ascent; conversely, descending air is warmed at the same rate—the *adiabatic lapse rate,* meaning that the lapse (the change of temperature) takes place within the parcel of air adiabatically (without loss or gain of heat from the outside). The dynamics of stability versus instability is an important concept in meteorology. It may be illustrated by the behavior of a hot-air balloon. Under conditions of instability, when the air outside the balloon is cooler than the air within it, the balloon will rise. When the temperatures are equalized, the upward dynamic push is lost and the balloon resists being displaced vertically. It has reached stability. In terms of weather, the swelling atop a cumulus cloud and the rising turret of a cumulonimbus, or thunderhead, are evidence of instability within the clouds. Such activity leads to an expansion of a cloud's height and frequently to the beginnings of some form of precipitation. On the other hand, stable air inhibits vertical motion (updrafts and downdrafts), and the formation of clouds. Existing weather and cloud conditions generally remain unchanged.

Results of Instability: Changes in the air's stability cause many weather processes. On a summer afternoon, for example, intense heating of the ground surface induces instability and causes air parcels to rise. This can lead to cumulus cloud formation and thunderstorm activity. At night, when the land surface cools, the air returns to a stable condition. Nighttime cooling of a stable air mass encourages fog formation in the early morning hours. Air may also become stable by moving over and coming in contact with a cold surface, such as ice, snow-covered ground, or cold water, which chills the air's lower levels. Conversely, air may become unstable when it moves over the warm surface of a lake or ocean.

Subsidence and Lifting: There are other ways in which air moves vertically. When high pressure prevails over an area, the air aloft tends to descend and spread out, or diverge, near the ground. In this process, known as *subsidence,* the sinking air is heated by the compression and clouds tend to dissipate and the sky clears.

The opposite of subsidence is *lifting,* whereby the temperature of an ascending column of air is lowered as it rises into cooler air. The air column's temperature eventually reaches its condensation level where invisible water vapor turns into visible water droplets. These appear as clouds and may eventually lead to precipitation.

Air may also be lifted through a process called orographic lifting, which occurs when a vertical barrier such as a mountain range or a front forces an advancing flow of air to ascend. In doing so, the air cools, often reaching its condensation temperature and forming clouds and perhaps precipitation. Often the rain falls only on the windward slope of the barrier, while on the leeward slope the descending air, cloudiness and precipitation are inhibited.

Air Masses: The atmosphere over North America is not homogeneous; at any given moment it is composed of several large bodies of air, each of which has different physical properties. Meteorologists call them air masses. They might also be called "oceans of air," but perhaps "large reservoirs" is a more descriptive term, because from time to time the atmospheric dams holding them in place break. Then the air flows away, usually toward the middle latitudes of the globe, to join the westerly airstream.

These airstreams journey thousands of miles. They can bring temporary tropical warmth to northern latitudes or shower conditions to desert areas. At any one time there may be three or four airstreams from different air-mass sources in motion over the United States and Canada, each covering hundreds of thousands of square miles laterally and extending aloft all the way to the tropopause, at the bottom of the stratosphere.

An air mass is technically defined as a large body of air exhibiting relatively uniform properties of heat, moisture, and stability. The surfaces of certain places in the Northern Hemisphere—such as the snow-covered tundras of Alaska and northwestern Canada and the warm waters of the Gulf of Mexico—act like giant air conditioners, imparting their native qualities of temperature, moisture, and cloudiness to the air above. Air tends to stagnate over these source regions; you can usually identify them on the weather map by the persistence of high atmospheric pressure for days at a time. In their birthplaces and early travels, air masses are usually closely associated with high-pressure, anticyclonic conditions.

Air masses are eventually set in motion by pressure changes induced by the general circulation of winds around the globe. When the restraining pressure lessens, the atmospheric dam dissolves. As the air mass moves away from the original source region, its temperature and moisture content are modified by different land surfaces it moves over and by mixing with other airstreams in a cyclonic storm system where cold and warm fronts contend.

A successful forecaster must be aware of the history of an airstream, considering the original conditions in the source region, tracing the effect of varied terrain in altering its physical properties during the journey, and judging the amount of mixing with other airstreams en route. For example, if you want to become acquainted with the current weather in Green Bay, Wisconsin, you will have to range to all points of the compass, studying the characteristics of air masses and terrain in diverse directions—the tundra of the Arctic, the vast surface and currents of the Pacific Ocean, the deserts of the Southwest and Mexico, and the warm waters of the Gulf of Mexico. Many different air masses combine to influence the weather and climate of northeastern Wisconsin.

Seven principal air masses affect weather in North America. They are described below and illustrated on the map on pages 36–37.

Arctic: The snow- and ice-covered terrain of central and northern Alaska, northern Canada, Greenland, the islands of the Arctic Ocean, and—especially—Siberia compose the source region of arctic air. Like its close relative, continental polar air, arctic air is very cold and very dry. In fact, the degree of coldness is the only property that distinguishes the two as different air masses. In summertime, when there is little temperature difference, arctic air and polar air are not distinguished on weather maps. In wintertime, the true

Air Masses

arctic air

continental polar air

maritime polar, Pacific air

continental tropical air

maritime tropical, Pacific air

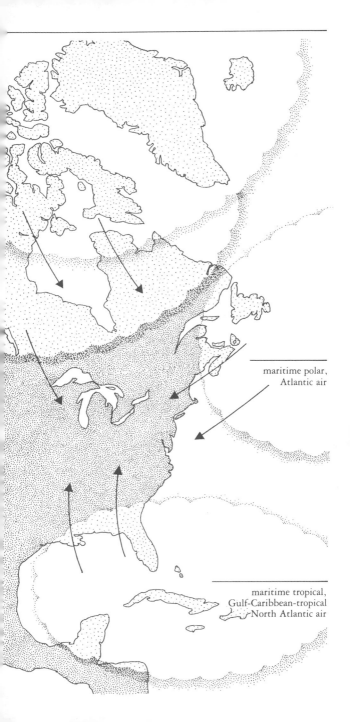

maritime polar,
Atlantic air

maritime tropical,
Gulf-Caribbean-tropical
North Atlantic air

arctic air mass of North America
usually lies over Alaska north of the
Brooks Range and over northern
Canada, poleward of a line running
from the lower Mackenzie Valley
southeast to southern Hudson Bay and
east to southern Labrador. Winter
temperatures often range down to
$-60°F$ ($-51°C$). After intense
radiation of heat skyward under clear
skies during the long winter night, ice
fog forms, especially in the vicinity of
settlements, lakes, or other moisture
sources. Snowfall is light, except where
airstreams pass over open water; then,
light to heavy snow showers may
develop on the lee shore. True arctic air
usually remains in Canada, making
only infrequent incursions into the
United States.

Continental Polar: The source region of continental polar
(cP) air is far-ranging, from central
Alaska southeast over the prairie
provinces of Canada, the Hudson Bay
when it is frozen, and east to Quebec
and Labrador. In spring and autumn,
the winter home of arctic air also
produces continental polar air.
Conditions are cold and dry within the
air mass, and the air is usually stable.
Clear nights prevail, and some fair-
weather cumulus clouds may develop in
daytime. When the airstreams are
heated from below, as they are in
passing over the warm, open waters of
the Great Lakes in autumn and early
winter, intense snow showers may
develop on the far shores and over the
Appalachian Mountain elevations.
These are known as lake-effect snows.
Polar air masses are capable of
penetrating deep into the southern
United States.

Maritime Polar, Pacific: The northern reaches of the North
Pacific Ocean constitute the immediate
source region of Pacific maritime polar
(mP) air. However, most of this mP air
originates over Siberia before crossing
the cool waters of the ocean south of the

Aleutian Island chain, where it acquires its cool and moist properties. In winter, it arrives on the North American coast as unstable air bearing copious moisture, which it delivers as rain or snow from the Alaskan panhandle to southern California. In summertime, Pacific maritime polar air brings a stable layer of cool temperatures and low-level cloudiness, known as the marine layer, to the immediate West Coast, making for foggy mornings and sunny afternoons and controlling the coastal weather of California. Inland in summertime, the maritime air is heated rapidly under intense sunshine, resulting in clearing skies in late morning and rapidly increasing temperatures.

Maritime Polar, Atlantic: Atlantic maritime polar air develops above the offshore waters of the northwestern Atlantic Ocean from the vicinity of Newfoundland south to Cape Cod. This air mass often has moved from central Canada and stalled along the coast, remaining partly over land and partly over sea, because it is blocked by a strong anticyclone in the Atlantic Ocean. Its airstreams are cool and moist, bringing low-level cloudiness, drizzle, and fog to the Atlantic seaboard as far south as Cape Hatteras. Springtime and early summer mark the most frequent appearances of Atlantic maritime air along the North American coast. Early summer heatwaves are often broken by the arrival of this cool airstream from the east or northeast. Boston's famous "east wind" results from a landward movement of polar maritime air, which has often hung offshore for several days before moving inland. Sometimes the airstream covers all the territory west to the Appalachians and south to the Virginia–North Carolina border.

Maritime Tropical, Atlantic: The tropical waters of the Gulf of Mexico, Caribbean Sea, and Atlantic Ocean between 10°N and 30°N constitute a source region for an important maritime tropical (mT) air mass. Maritime air helps to maintain the global temperature balance by transporting vast quantities of heat from tropical to temperate latitudes. These northward-moving airstreams are warm and moist, with a tendency toward instability. When drawn into cyclonic storms moving eastward over the United States, they supply the warm sectors of those systems with moisture for precipitation, either in shower or steady form. Their influence ranges west to the Rocky Mountains and north to southern Canada.

In summer, maritime tropical air dominates the weather pattern of the eastern half of the United States, bringing heat and humidity. The relatively heavy annual rainfall of the southern states results from their proximity to the moisture from the Gulf of Mexico.

In winter, maritime tropical air is sometimes drawn northward into storm systems, supplying the moisture for heavy snowfalls along the Atlantic seaboard. When this air is excluded by strong westerly flow from the southern and eastern United States, drought conditions develop and threaten agricultural production and urban water supplies.

Maritime Tropical, Pacific: The tropical portions of the Pacific Ocean, from the west coast of Mexico and Central America to Hawaii and beyond, are the source region of Pacific maritime tropical (mT) air. For most of North America, this air is not of major importance, but for the western and southwestern states it has considerable significance. When Pacific lows strike southern California and Baja California, they carry tropical Pacific air onto the continent, dropping their moisture over the mountain ranges and desert areas of the Southwest. The summer showers that break out in the warm months over the southwestern states can be traced mainly to airstreams from the tropical Pacific Ocean. Errant tropical storms from the west coast of Central America and Mexico sometimes transport warm, moist tropical air up the Gulf of California to the southwestern states.

Continental Tropical: In summertime, the deserts of northern Mexico and the southwestern United States produce a very dry, intensely hot air mass that occasionally moves northeast into the Great Plains. Though it is usually unstable, this continental tropical (cT) air mass has such a low moisture content that little cloudiness or precipitation results. If the air mass persists for several weeks, drought conditions may develop in the southwestern states and across the plains.

Wind: Meteorologists define wind as air in horizontal motion relative to the surface of the earth. The vertical components of atmospheric motion are relatively small, especially near the surface of the earth, and are usually identified as ascending or descending currents, not wind.

The general circulation of air around the globe is a response to differences in atmospheric pressure caused by three main factors: thermal differences in solar heating with respect to latitude; the influence of the rotation of the earth; and the unequal distribution of land and oceans around the globe.

Circulation Cells: Temperature differences resulting from unequal receipt of solar radiation in various latitudes account for the pressure differences that drive the general circulation. The Northern Hemisphere is divided into three circulation regimes, or cells: an equatorial-tropical cell, a middle-latitudes cell, and a weak polar cell. (The Southern Hemisphere has similar, "mirror-image" regimes, with which we are less concerned in this guide.)

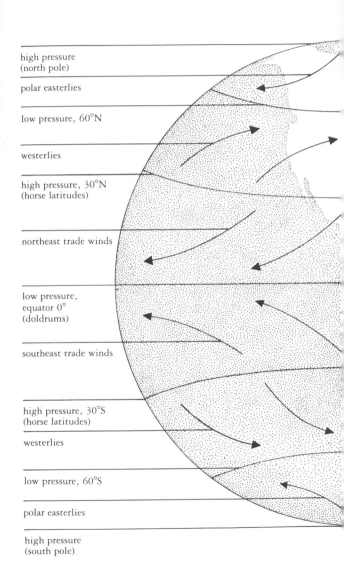

high pressure
(north pole)

polar easterlies

low pressure, 60°N

westerlies

high pressure, 30°N
(horse latitudes)

northeast trade winds

low pressure,
equator 0°
(doldrums)

southeast trade winds

high pressure, 30°S
(horse latitudes)

westerlies

low pressure, 60°S

polar easterlies

high pressure
(south pole)

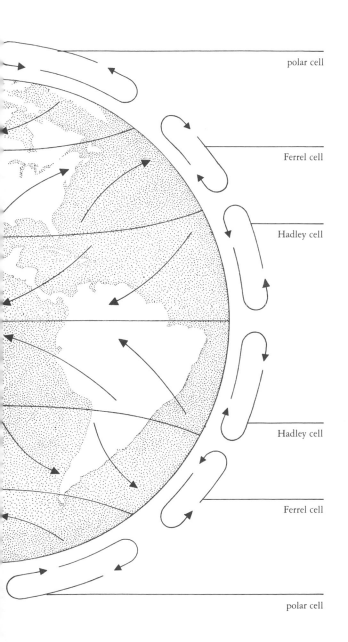

polar cell

Ferrel cell

Hadley cell

Hadley cell

Ferrel cell

polar cell

Air near the equator tends to rise as a result of strong heating of the surface. Once aloft, it flows northeastward over the tropics but descends again near 30°N as a warm, drying air current. (The desert regions in subtropical latitudes are the results of such flow.) The air at the surface returning southward is deflected toward the west as trade winds, which blow from the northeast. They meet southeast winds from the Southern Hemisphere in the equatorial region, forming a convergence zone where cloudiness and heavy rainfall prevail. These circulation regimes have long been known as Hadley cells, after George Hadley, the early eighteenth-century English physicist who first proposed the theory of their cause and existence.

Mariners have given names to the zones of calms or light shifting winds at the borders of the Hadley cell. The equatorial zone between 10°N and 10°S is known as the Doldrums, taken from our word for a spell of listlessness. The zone between 30°N and 35°N is called the Horse Latitudes, so called because of the practice of throwing animals overboard when ships were becalmed for lengthy periods of time and ran short of food and water.

A separate circulation exists between 30°N and 60°N (with, of course, a mirror image between 30°S and 60°S). Upper-air winds in this zone usually blow from a generally westerly direction. The westerly flow of the Northern Hemisphere cell is regularly interrupted by local storm circulations over the United States and southern Canada. These storms, however, often become embedded in the westerly flow and are carried eastward over the Pacific Ocean, North America, and the Atlantic Ocean. This middle-latitude circulation is named the Ferrel cell, after a nineteenth-century American meteorologist.

A weak cell with surface easterly winds occupies the arctic region above 60°N. An outstanding characteristic of the Arctic circulation is the occasional outbreak of polar or arctic air that brings frigid, low-level airstreams into the middle-latitude zone. This transfer contributes greatly to maintaining the global temperature balance.

The Prevailing Westerlies: The major airflow over the United States is from slightly south of west to slightly north of west. To understand more clearly the reason for the westerly winds, consider the relationship between horizontal pressure differences and horizontal temperature differences. Pressure decreases more rapidly with height in cold air than in warm air. Because cold air is denser than warm air, it is more compressed in the vertical, and exerts more pressure at the earth's surface. Conversely, if the pressures of a warm column and a cold column of air are equal at the surface, then at any higher level the pressure would be greater in the warm column than in the cold column. In the troposphere, the warmest air is located near the equator, and temperatures decrease toward the poles. Therefore, pressures at higher altitudes are greater in the equatorial zone, causing a flow toward the poles. But the eastward rotation of the earth causes a deflection of air—to the right in the Northern Hemisphere and to the left in the Southern Hemisphere. This phenomenon is known as the Coriolis force, named for its discoverer, Gaspard Coriolis, a French civil engineer of the nineteenth century.

The Coriolis force is zero at the equator and increases steadily with increasing latitude, thus deflecting air flowing toward the poles in both hemispheres toward the east, and reinforcing the west-to-east winds that prevail over the middle latitudes. To visualize the

The Jet Streams

Waves in the Jet

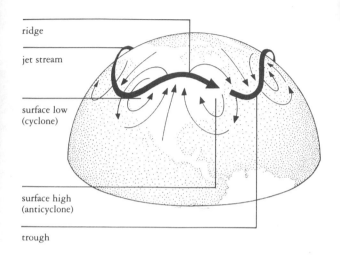

ridge

jet stream

surface low
(cyclone)

surface high
(anticyclone)

trough

The Jet Stream in Winter

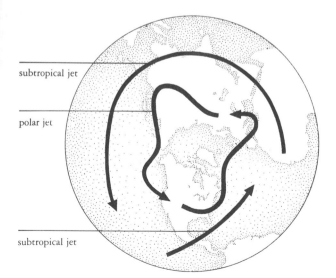

subtropical jet

polar jet

subtropical jet

Coriolis force, imagine the following: A rocket was launched at the north pole on a straight path south for an hour's flight. During the course of an hour, the earth rotates on its axis 15 degrees (1/24 of the full circle, 360 degrees, that it spins each day). If the rocket follows a course due south while the earth spins, the craft will land 15 degrees west of the place where it was launched. The earth's rotation gives the rocket trajectory a curved path to the right (west) when plotted on the earth's surface. A north wind, starting from the North Pole, would be a northeast wind upon reaching the equatorial zone since the deflecting force is always directed at right angles to the direction of air flow. (See the illustration on page 28.)

The Jet Stream: Vertical pressure differences increase with altitude in the troposphere because a large north–south temperature gradient exists there. Thus the speed of the westerly air flow increases to reach a maximum just below the tropopause. The high-speed flow aloft is called the *jet stream,* a relatively narrow ribbon of very fast westerlies over the middle latitudes at an altitude of 6–8 miles (about 10–13 km) above the earth's surface. Speeds can reach an extreme of more than 200 miles per hour (about 322 kph), though they generally average about 80 miles per hour (about 129 kph) in winter and 40 miles per hour (about 64 kph) in summer. The importance of the jet stream lies in its position as the meeting place of cold airstreams from the north and warm airstreams from the south. Storms tend to generate in the boundary zone below the jet, known as a front, and are steered on an eastward course by the jet stream above. The jet seldom flows in a straight path, instead meandering north and south in sinuous curves as it makes its global journey. (See the drawing on page 46.)

In winter, two jets can often be identified across North America. The polar jet, normally found in the vicinity of the United States–Canada border, separates air of polar origin from air that has been conditioned in middle latitudes. To the south of the polar jet, the subtropical jet is found over the southern parts of North America. The subtropical jet marks the northern limit of the Hadley cell, and frequently arrives over northern Mexico in a strong southwesterly flow from the Pacific Ocean and follows a northeastward path over Texas and the southern states. In summer, as the pressure zones shift northward and tropical air occupies the midcontinent, only the polar jet remains on weather maps, crossing the middle latitudes of the Pacific Ocean and the southern provinces of Canada.

Waves in the Jet: The upper-air flow tends to encourage wave motion in the westerly airstream in much the same manner that air movement over the ocean causes the formation of waves on the water's surface. Air waves take the form of atmospheric troughs and ridges, which occasionally force the normally westerly flow to shift north or south. The portion of the jet-stream wave that bows to the north is called a ridge, and the part that dips to the south is a trough. The upper-air flow pattern appears on a map as a series of long waves, in which four to six sets of troughs and ridges succeed one another in a somewhat regular manner around the globe. (See the drawing on page 46.) It is the constant changes in speed, direction, and location of wave formations in the upper-air flow that bring such variety to the weather patterns of the United States and Canada. It isn't only weekly, monthly, and seasonal trends of temperature and precipitation that follow the dictates of the westerlies; daily and even hourly changes of weather are also determined by the properties of the air masses that the upper-air flow brings to our vicinity. The fundamental dynamics of upper-air circulation and its wave structure have been under intensive study during the past 40 years. Assisted by computers, researchers have learned enough about the flow patterns to devise fairly accurate models of normal atmospheric behavior. Supplied with current data, computers can prepare outlooks for daily, weekly, and monthly trends in circulation, providing the basis of area and regional temperature and precipitation forecasts.

Traveling Cyclones and Anticyclones: Although jet streams pass many miles over our heads, they breed storms that directly affect our lives on earth. The frontal boundary that lies beneath the jet stream takes on the same sinuous pattern of the jet, and as the waves aloft shift eastward, so do the waves on the front. This shift results in alternating cold and warm air masses. Ridges (the northward-flowing portions of the jet-stream wave) are generally associated with warm-air masses, and beneath the northward bulge of the jet stream there is usually warm air intruding from the tropics. The leading, northernmost edge of this warm air is known as a "warm front." Troughs, on the other hand, overlie cold-air masses escaping south from the arctic. The leading, southernmost edge of cold air is a "cold front."

The meeting of these different types of air generates weather disturbances. Warm air is lighter, and usually contains more water vapor, than cold air. Thus warm fronts and cold fronts have different characteristics. Along warm fronts, the warm air overruns the cold, bringing thickening and lowering clouds followed by steady precipitation. With cold fronts, the heavier cold air cuts under the lighter warm air, forcing the warm air upward, often in a fairly

Fronts

Cold Front

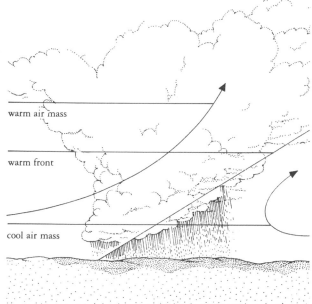

cold front

warm air mass

cool air mass

Warm Front

warm air mass

warm front

cool air mass

narrow line, setting off brief but heavy precipitation in showers, squalls, and thunderstorms.

Waves on the frontal boundary between arctic and tropical air masses are not stable, and they have a tendency to develop into cyclones. Cyclones generally develop west of the warm, southerly airstream beneath the ridge and east of the cold northerly flow beneath the trough. Meanwhile, anticyclones (or surface highs) are found beneath the southeastward-flowing part of the jet. The counterclockwise circulation around the cyclone and clockwise circulation around the anticyclone pull the cold air farther south and the warm air farther north, and as a result the wave on the jet stream grows in its north–south amplitude. However, the mixing of arctic and tropical air masses reduces the temperature contrast between north and south, and eventually the cyclone, anticyclone, and jet-stream waves all weaken and dissipate. As the arctic and tropical air masses begin to reestablish themselves in their usual locations, the jet stream builds up once again and the cycle repeats itself.

Storm Tracks: Cyclones and anticyclones that form on the front beneath the polar jet often travel considerable distances (thousands of miles) as they live out their life cycles. Most traveling cyclones and anticyclones move in the direction of the overlying jet stream—that is, cyclones move northeast and anticyclones southeast. For cyclones, these paths are known as "storm tracks." East (or ahead) of a traveling cyclone, a northbound warm front marks the limit of the tropical southerly airstream, and for those in the path of the cyclone rising temperatures often precede the stormy weather (often called the "warm before the storm"). To the west of (or behind) the cyclone,

Large-Scale Airflow over North America

Semipermanent Highs and Lows

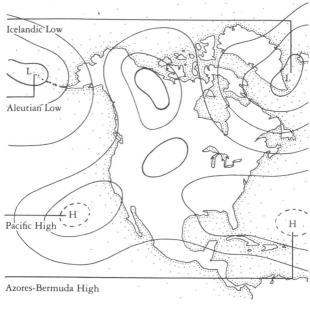

Icelandic Low

Aleutian Low

L

Pacific High

H

H

Azores-Bermuda High

Straight Westerly Airflow
(No dominant storm track)

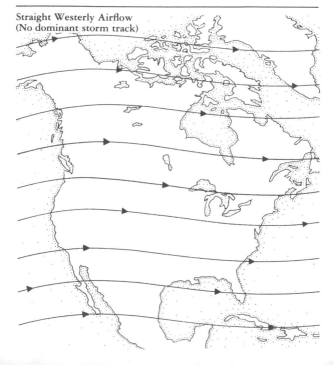

cold northerly winds follow the southward advance of a cold front, and typically the passage of a cyclone and the end of precipitation bring clearing and colder weather. As the cyclone moves away, the anticyclone enters the scene. The anticyclone, or "high," ushers in clear skies that may last for several days. As the anticyclone passes overhead and moves east, the winds turn southerly in advance of the next cyclone. This is the usual sequence of weather accompanying the passage of traveling cyclones and anticyclones in middle latitudes. But remember that in meteorology the "usual" generally means the average of many events. Thus cyclones and anticyclones may approach from any point of the compass; they may stall overhead, or suddenly strengthen or weaken when least expected. Even with the infinite possibilities of behavior exhibited by highs and lows, there are places where cyclones are more likely to occur, and other places where anticyclones are more likely. Although cyclones and anticyclones may not be present at these locations every day, they can be persistent enough to be called "semipermanent highs and lows."

Semipermanent Highs and Lows: On a typical weather map of the Western Hemisphere, you will usually see two large ridges of high pressure near North America. They are usually found over the middle latitudes of the Atlantic and Pacific oceans, and their peripheries lie adjacent to the east and west coastlines of the United States. These semipermanent highs are of such considerable vertical extent that they appear on both surface and upper-air weather maps. (See page 52.)

Highs: The Pacific High, lying roughly north and northeast of the Hawaiian Islands, is important in steering weather across the ocean. It is a migratory body, moving northward in summer and retreating toward the tropics in winter. The flow lines of the westerlies arch northward around the Pacific High, passing below the Aleutian Islands and through the Gulf of Alaska before heading for lower latitudes on the North American coast. Storm centers born off the coasts of Japan and Siberia are steered around the Pacific High into the Aleutian Low. Frequently the Aleutian Low regenerates old storms or creates new ones that strike at the North American continent.

In the Atlantic Ocean a similar pattern exists, although on a smaller scale. A large area of semipermanent high pressure extends from the vicinity of Bermuda eastward to the Azores and on to the coast of Spain. The main area is known as the Azores High, and its western extension is called the Bermuda High; the all-inclusive term used by American forecasters is Azores-Bermuda High. Like its Pacific cousin, this high is a migratory body, and its shifts northward and southward, westward and eastward, help determine the tracks of storms in the western Atlantic Ocean. Even the West Indian hurricane, monster that it is, cannot break through the atmospheric block provided by a vigorous and well-established Bermuda High. Such a visitor from the tropics is often shunted westward into the Caribbean Sea and the Gulf of Mexico.

Lows: Complementing the two semi-permanent high-pressure ridges are two notable areas of low pressure, where trough conditions prevail through most of the year, but are strongest in the winter. These lows have a great influence on the development and progress of most storms in the middle and high latitudes of the Northern Hemisphere. The one of major importance to weather in the United

States and Canada is the Aleutian Low, which occupies an oceanic position close to the Aleutian Islands, between Siberia and Alaska. Here cold airstreams from the Arctic region and warm airstreams from the central Pacific Ocean confront each other and create active weather disturbances.

The Aleutian Low is the major storm factory of the world. Small circulatory wind systems, spinning away from the parent low, become vigorous cyclonic disturbances and sweep southeast across the Alaskan panhandle into British Columbia and the northwestern United States. Sometimes the cyclonic systems come in a series of three or four, at intervals of 24–36 hours, to bring stormy conditions to the entire West for a week or more.

After crossing the western mountains, the systems tend to turn northeast across the continent and head for the Icelandic Low, the second semipermanent low-pressure area of major importance. The Icelandic Low covers a broad expanse of the North Atlantic Ocean, with the lowest pressures found between Iceland and the southern tip of Greenland. It serves as a meteorological magnet, attracting traveling cyclonic systems from the western Atlantic basin. Here, some systems regenerate or new storms form to strike eastward at northwestern Europe.

Airflow
Patterns:
The dynamics of airflow require that troughs follow ridges in an endless succession around the globe. Air approaching a ridge streams northeast until it reaches the crest, then turns toward the southeast while descending the eastern flank of the ridge. The air flow continues toward the southeast into the next trough downstream. Upon reaching the bottom of the trough, a turn to the northeast takes place as the flow ascends the western

flank of the next ridge.

Depending on the location of the semipermanent troughs and ridges and on the position and strength of migrating cyclones and anticyclones, different airflow patterns can prevail across North America during the course of a week, month, or year. The five different types of airflow patterns may continue for periods varying from several days to several weeks.

Straight Westerly Flow: The simplest pattern features straight west-to-east flow across North America with little wave action. (See the map on page 52.) The flow is well established in breadth and extends to great heights. Wind speeds are usually high, and weather disturbances are carried rapidly from west to east on a track across southern Canada. The journey across the continent takes about three days in wintertime and longer in the warm season. Slight ridges and troughs may appear in the wave pattern, but they are relatively small; straight west-to-east flow is the rule. The airflow originates over the Pacific Ocean, bringing moderately cool, moist air over the areas west of the Rocky Mountains. The airstreams lose much of their moisture in rising over the mountains, and they descend onto the Great Plains as a mild, dry airflow. In the vast interior of the continent, the modified Pacific air may mix briefly with airstreams from Canada or the Gulf of Mexico, but the flow from these regions is soon cut off by a renewed impulse from the westerlies. Only weak frontal activity and brief periods of precipitation will usually prevail from the Great Plains and Mississippi Valley eastward to the Atlantic seaboard.

Under these conditions, the Pacific and Bermuda highs are well developed and extended along the lines of latitude, overlapping the continent in California on the West Coast and the Carolinas on the East Coast. The Aleutian and Icelandic lows are also well developed and centered north of their usual wintertime positions, tending to spread out on a lateral axis and thus cover most of Canada with low pressure. Cyclonic disturbances enter the continent over British Columbia, or they may originate over the northern Rockies, and then move eastward close to the boundary of the United States and Canada. Their tracks eventually trend northeast from the Great Lakes to Labrador, since a direct easterly course is blocked by the expanded influence of the Bermuda High over the Atlantic Ocean.

After a period of strong flow from the west lasting from one to four weeks, the speed of the wind decreases and westerly control of the weather diminishes. The upper-level westerlies and underlying front separate arctic air to the north from tropical air to the south, and as long as there are only weak troughs in the westerly flow there is little mixing of the opposing air masses.

In the absence of mixing, the temperature contrast between the arctic and tropical air masses increases, thereby increasing the strength of the westerlies and setting the stage for a change in the large-scale airflow pattern. Enormous north–south undulations develop in the westerly flow, and the straight west-to-east airflow is replaced by a pattern dominated by troughs and ridges. This pattern permits streams of warm air from the tropical stretches of the Pacific or Gulf of Mexico to mix in the circulation pattern at the same time that cold polar air from Alaska or Canada is drawn southward into the United States. Cold fronts sweep down from Canada and may penetrate all the way to Texas and the Gulf of Mexico; in turn, warm fronts may take form in the South and bring balmy breezes as far north as southern Canada. Fronts are sharp, contrasts of air mass are marked, and disturbances flourish along the meeting zone of the cold and warm airstreams.

Ridges and troughs under this regime are always in a state of unrest, like ocean waves, with constant changes in position and size. Nevertheless, meteorologists have been able to reduce the infinite variety of trough and ridge formations to the four main types described below. Although these are oversimplifications of the always complex weather situation, they are very valuable in helping to illustrate the controls the upper-air flow exerts on our weather.

Western Trough-Eastern Ridge: The main feature of this airflow pattern (see the map on page 60) is a marked area of low pressure in the form of a trough lying to the west of the Rocky Mountains, over the Intermountain region or the Pacific states. To the east of the Mississippi Valley a large ridge dominates, with its crest over the Appalachians or the Atlantic states. This arrangement permits a strong, persistent flow of northwest winds down the western side of the trough, bringing cold air from the north Pacific Ocean or Alaska to the states west of the Continental Divide. In the East, airflow is from the southwest on the western flank of the eastern ridge, and either tropical air or polar air (modified by a stay over the United States) brings mild conditions far northward. Cold weather is the rule in the West; warm conditions prevail in the East.

The troughs and ridges in the upper-air flow are not stationary; they tend to shift eastward, guiding the traveling low- and high-pressure systems at the surface. They also exhibit a general tendency to recur and reestablish themselves, enabling an educated observer to make broad generalizations about the accompanying surface weather.

Under the western trough–eastern ridge pattern, cyclonic storm systems from the Aleutian Low move into the West Coast from British Columbia and the Pacific Northwest to as far south as California, delivering heavy precipitation to western coastal and mountain localities. Farther east, there is a marked tendency for storms to develop or regenerate in the lee of the Rocky Mountains, mainly in Colorado and the panhandle regions of Texas and Oklahoma, and these disturbances move northeast in the broad southwesterly flow on the eastern side of the trough. Once the moisture source in the Gulf of Mexico has been tapped, such storms produce heavy rain or snow in the Northern Plains, upper Mississippi Valley, and Ohio Valley. Storminess east of the Mississippi River is generally confined to the Middle West, lower Great Lakes region, and northern New England. Anticyclonic conditions in the ridge keep the Southeast and the Middle Atlantic states warm and dry unless a cold front sweeps southeastward and causes a brief period of showers.

Western Ridge-Eastern Trough: This type of airflow is distinguished by a strong ridge over the Intermountain and Rocky Mountain regions, with the north–south axis of the low-pressure trough east of the Mississippi River, either along the western slope of the Appalachians or on the Atlantic coast. (See the map on page 60.) Cyclonic systems from the Pacific Ocean may move toward the Pacific Coast, but they are shunted north into the panhandle of Alaska, northern British Columbia, and the Yukon as they are steered around the Intermountain High, which forms in the core of the western ridge. Once the disturbances move around the northern periphery of the ridge and into the northwesterly

Airflow Patterns

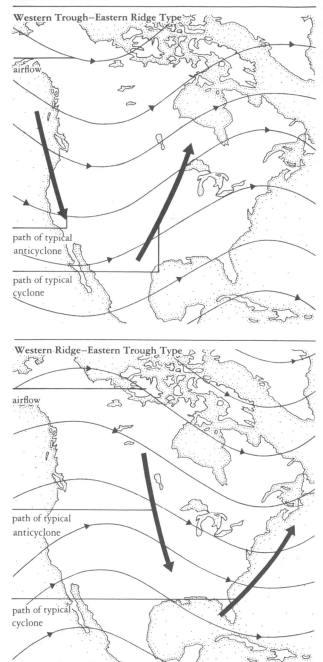

Western Trough–Eastern Ridge Type

airflow

path of typical
anticyclone

path of typical
cyclone

Western Ridge–Eastern Trough Type

airflow

path of typical
anticyclone

path of typical
cyclone

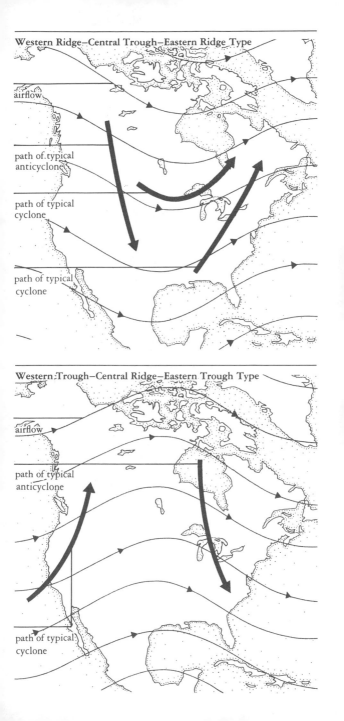

Western Ridge–Central Trough–Eastern Ridge Type

airflow

path of typical
anticyclone

path of typical
cyclone

path of typical
cyclone

Western Trough–Central Ridge–Eastern Trough Type

airflow

path of typical
anticyclone

path of typical
cyclone

flow to the east of the ridge line, they are carried rapidly southeast, usually followed by outbreaks of cold, polar air from northwestern Canada and accompanied by light precipitation. The Pacific Coast and the Intermountain region experience warm daytime temperatures and normally clear skies; cloudy skies and cold conditions prevail over the eastern two-thirds of the country. Even the Gulf Coast and Florida can experience a freeze in wintertime as a result of a northerly airflow.

As a cold front enters the trough along the South Atlantic Coast and cold polar air passes over the warm waters of the Gulf of Mexico or the Gulf Stream off Georgia and the Carolinas, storms can develop quickly. They move northeast along the coast, delivering driving snow in winter and cold rain in summer—the famous northeasters (a storm wind flowing from the northeast) of the Atlantic seaboard. Cold air from central and eastern Canada drawn into the circulation gives the Middle Atlantic and New England states their most severe cold waves.

Western Ridge- *Central Trough-* *Eastern Ridge:* This third trough-ridge pattern (illustrated by the map on page 61) is a more complicated version of the two described previously. Its main feature is a distinct low-pressure axis running south–southwest through the Mississippi Valley to Texas and Mexico. There are marked pressure ridges to the east and west of the trough, making for sharp temperature contrasts across the country. Cyclonic systems move inland from the North Pacific around the northern periphery of the western ridge, dip southward across the Great Plains, and head east and northeast over the Great Lakes and down the Saint Lawrence Valley. A notable feature of this alignment is the tendency for vigorous secondary storms to form in the southern end of the central trough over southeastern Colorado, Texas, or the western Gulf of Mexico at the same time that the primary storm center is passing over the Great Lakes. The secondaries move rapidly northeast toward the Ohio Valley, often becoming the main storm center. They bring widespread stormy weather to all the Midwest and the Northeast, with heavy rainfall, thunderstorms, and tornadoes to the east of the storm track and, in winter, frozen precipitation and sometimes blizzard conditions to the west and north.

Western Trough-Central Ridge-Eastern Trough: The final major airflow pattern features cyclonic activity on both coasts, an anticyclonic regime of clear skies and below-normal temperatures over the Rocky Mountains, and above-normal temperatures in the vast Mississippi Valley. (See the map on page 61.) The main cyclonic activity brings North Pacific disturbances over the West Coast south of Washington. In the absence of an Intermountain High, these disturbances cross that region on a southeast track and attain the upper watershed of the Rio Grande River, then fade from the weather map before reaching the shores of the Gulf of Mexico. An abundance of precipitation may occur over the mountains of Oregon and northern California, but normal amounts fall over the Intermountain region and very little over the southern Rockies.

Sometimes, small disturbances known as waves develop along the cold fronts that have reached the Gulf of Mexico or the Gulf Stream off the Georgia-Carolina coast. If the path of the newly created storm center lies close enough to shore, the shield of precipitation will cover the Middle Atlantic and New England coastal areas. Such storm conditions, which move northeastward, usually develop in midwinter, but they have occurred as early as September and as late as June.

THE SEASONS

Our ever-changing seasons result from the fact that the earth's axis (the line running between the North and South Poles through the center of the earth, and around which the earth rotates) is not perpendicular to the plane of the earth's orbit around the sun. The axis is actually tilted by 23° 26', and the direction of the tilt is such that on or about June 21 (the summer solstice) the North Pole is tilted 23° 26' *toward* the sun, and six months later is tilted 23° 26' *away from* the sun. As a result, as the earth moves around the sun in its annual orbit, solar rays arrive on the planet at changing angles. During half the year the sun's rays fall most directly on the Southern Hemisphere, and during the other half of the year they strike the Northern Hemisphere most directly. If the axis were perpendicular, we would have no seasons, but would experience the same range of weather every day of the year. (See the illustration on page 68.)

The Astronomical Seasons: From our perspective on earth, the sun's daily track across the sky reaches varying angles above the horizon, achieving its highest angle in the Northern Hemisphere at noon on June 21. (At 23° 26' North latitude—the same angle as the axis's tilt—the sun passes directly overhead on this date.) On the same day, the sun spends more hours above the horizon than on any other date. These two factors, high sun angle and length of day—factors that define the astronomical seasons—allow more solar energy to heat the ground than on any other day of the year. Conversely, on or about December 21, the solar heating reaches its minimum in the Northern Hemisphere, while it peaks in the Southern Hemisphere. Astronomically, these dates, June 21 and December 21, mark the beginnings of the summer and winter seasons, respectively, in the Northern Hemisphere. Approximately midway between these two dates, around March 20 and September 22, we mark the start of the Northern Hemisphere's astronomical spring and autumn. The earth's axis is perpendicular to the earth-sun line, and neither of the earth's poles is pointed toward or away from the sun. As seen from earth, the noontime sun passes directly overhead on the equator, and the two hemispheres are heated equally. The starting dates of each of the four astronomical seasons sometimes occur a day later than the dates noted above because the earth doesn't conform exactly to our calendar, which is based on a 365-day year. The earth actually takes 365.24 days, or nearly six hours more than 365 days, to complete its circuit around the sun. As a result, every year the first moment of each season is six hours later by our calendars and clocks than it was the previous year. After four years, the seasons begin 24 hours, or one day, later. For this reason, we add an extra day to the calendar at this point, creating a 366-day leap year, and the seasons go back to their original starting dates.

Length: The lengths of the seasons are not exactly equal, because the path of the earth around the sun is not a circle, but an ellipse. In 1609, Johannes Kepler observed that varying gravitation causes the earth to move faster when it is closest to the sun (near its perihelion) and slower when it is farther away (around its aphelion). Since perihelion occurs in early January, the quickened orbit makes the autumn and winter seasons in the Northern Hemisphere slightly shorter in astronomical terms than spring and summer. The precise durations of the northern seasons are: 92.76 days of spring, 93.65 days of summer, 89.84 days of autumn, and 88.99 days of winter, totaling 365.24 days in a year. Even though the earth is closest to the sun in January, the sunlight striking the earth is only 6 percent stronger than at July's aphelion—not enough to offset the much larger effect of the tilted axis. And despite the timing of the perihelion, January is still colder than July in the Northern Hemisphere.

The Meteorological Seasons:
Meteorological seasons differ from the astronomical seasons described above in that they are defined by temperature rather than by the earth's position. Because our temperatures are affected by a phenomenon known as heat lag, the starting dates of meteorological seasons are quite different from those of astronomical seasons. On earth, the hottest time of year is, on the average, about a month after the first day of summer as noted on the calendar, and the coldest days occur about a month after the onset of winter. Even though the solar heating of the Northern Hemisphere declines after June 21, the heating remains relatively intense for several more weeks and continues to warm the masses of ground, air, and water. Since the temperature peak lags behind the solar-heating peak, this phenomenon has been named heat lag; it also explains why daytime

The Seasons in the Northern Hemisphere

The Astronomical Seasons/Orbit of the Earth

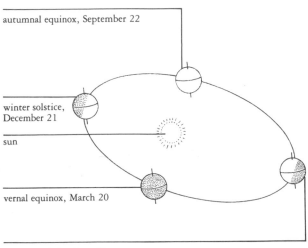

autumnal equinox, September 22

winter solstice,
December 21

sun

vernal equinox, March 20

summer solstice, June 21

The Meteorological Seasons/Annual Cycle of Temperature

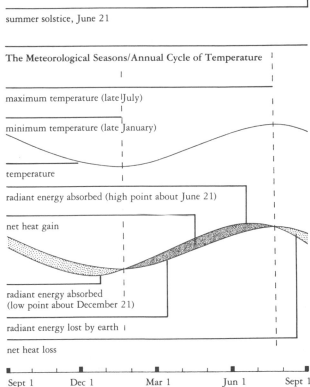

maximum temperature (late July)

minimum temperature (late January)

temperature

radiant energy absorbed (high point about June 21)

net heat gain

radiant energy absorbed
(low point about December 21)

radiant energy lost by earth

net heat loss

Sept 1   Dec 1   Mar 1   Jun 1   Sept 1

temperatures peak around 3:00 P.M. rather than at high noon. (See the illustration on page 68.) Because oceans and lakes heat and cool more slowly than does land, seasonal extreme temperatures usually occur even later over water and coastal areas than they do inland. For example, along coastal California the warmest months are August and September, rather than July, as is typical of inland locations. There are many different definitions for each meteorological season, depending on interests and needs. For example, corn farmers and Great Lakes ship captains have very different ideas as to when spring begins, and those who live in areas where no snow ever falls have a different measure of when winter unfolds. We will now examine some of the more common definitions of the meteorological seasons and show how their timing varies across North America.

Winter's Start: Meteorological winter has different definitions in different parts of the United States, and in this sense can be said to begin on different dates. The conventional commencement of winter is December 1, with the season continuing through February. By December 21 or 22, the date for the start of astronomical winter, most North Americans consider themselves well into their meteorological winter season.

In much of the West, winter is defined as the coldest quarter of the year, and so begins during the last week of November. In the Northeast, it starts the first week of December. From North Carolina to Idaho, and also in an area close to the Pacific coast, winter's start is on or close to December 1. Hence, for much of the nation's population the first day of winter agrees with the traditional calendar date. In the colder parts of the country, winter has begun when the daily average temperature, averaged over many years, falls below freezing. On this basis, winter starts during the last ten days of November for a large part of the northern area of the country, and as early as the first ten days of November in North Dakota and northwest Minnesota. It does not start until December 15 in a wide zone extending from eastern Massachusetts to Arizona. Much of the South and some Pacific coastal zones do not have daily normal temperatures below 32°F (0°C) at any time and thus do not have meteorological winter by this definition.

For some, the coming of snow indicates the arrival of winter. The first snowflakes fall in October or early November in northern and mountainous sections of the country, and as early as September in the central and northern Rockies. In the populated urban zone from New England to the Mississippi River, the first snowfalls occur in late November or early December. However, substantial accumulations (3 or more inches of snow) do not normally occur until mid-December in this region.

Advent of Spring:

When the year is divided into quarters of equal length, spring is the quarter intervening between winter and summer. If calendar months are used, spring includes the first day of March to the last day of May. If the astronomical migration of the sun is employed, spring starts with the vernal equinox, on March 20, 21, or 22, when the sun crosses the plane of the earth's equator on its journey from the Southern to the Northern Hemisphere. At that time days and nights are of equal duration.

Spring begins when winter ends, and if winter is considered the coldest quarter of the year, spring follows it in late

February in much of the South and far West; spring starts around March 1 for the zone extending from North Carolina through Missouri and Kansas to Idaho; and anytime from March 4 to 12 in the Northeast.

In the northern states, most people think spring has begun when average daily temperatures are above freezing. By this definition, spring generally starts about March 1 in the area extending from Boston and New York to Pittsburgh, Omaha, and northwest to the border of Idaho and Canada. The date is March 15 for those who live anywhere from Portland, Maine, through central New York State and westward to Detroit, Milwaukee, and Sioux City.

Some people think that the threshold of spring should be the date when the daily normal temperature reaches 50°F (10°C), about which time most plant life has revived from its winter dormancy. Such temperatures occur about February 1 in the Deep South, but not until April 1 in the area from Virginia to Kansas. It is much later, about May 1, in New England, northern New York, along the Canadian border from Michigan to Montana, and southward into the Rocky Mountains.

Summer Arrives: The summer quarter of the calendar year begins on June 1 and extends through July and August. June 20 to 22, the time of the summer solstice, when the sun's rays reach the Tropic of Cancer (23° 26′ North) and begin their southward retreat, is the beginning of astronomical summer.

Meteorological summer has several definitions. If it is based on the hottest quarter of the year, it begins in late May in much of the South and far West, about June 1 in a zone extending from North Carolina to Kansas to Idaho, and about seven days later in the zone that includes the land from New York State to North Dakota. Summer can also be defined as the season when the average daily temperature reaches 68°F (20°C). This normally happens in the Deep South by May 1, by June 1 for a zone from Maryland to Kansas, and not until June 15 in much of the West and Northwest. Some mountain locations in the Northeast and West, along with some coastal areas from northern California to Washington, never see summer at all, by this criterion.

Summer Ends And Fall Begins: Using the definition of summer as the quarter of the year having the highest average temperatures, summer's end takes place in the last week of August in the southern half of the United States and west of the Rocky Mountains. A notable exception is the immediate coastal regions of northern California, where the warmest quarter does not end until some time around September 30, or even until late October in San Francisco. The first week of September brings the end of summer by this definition in most of the remainder of the country, except in northern New England, where the date lags to about September 10. If summer is defined as the season having an average daily temperature above 68°F (20°C), it normally ends in August in the northern states, in September in the middle zone of the country, and in October in most of the southern states. On the peninsula of Florida and in southern Texas, the date occurs in November.

The Seasons In Alaska And Canada: North of the 49th parallel, which approximately defines the U.S.–Canada border west of the Great Lakes, the seasons revolve around the annual freeze–thaw cycle. A long winter of snow-covered terrain and cold prevails everywhere, except along the immediate Pacific coast. The spring thaw comes gradually: Snow and ice melt from land, lakes, streams, and seashores. The summers are generally warm, except along the northern Arctic coast, but are followed by the rapid freeze-up of autumn.

Winter: The start of winter is marked by meteorologists as the date when the air temperature drops to a consistent mean of 32°F (0°C). Usually about two weeks later, lake freeze-overs will occur, though some of the larger lakes may take longer, and the maximum mean temperature will fall to freezing. From then until spring, the temperature usually remains below freezing. By December 1, all is solid snow or ice for the rest of the winter.

Spring: The spring break-up comes when air temperatures regularly rise above a mean of 32°F (0°C). The persistent thaw enters the Okanagan section of British Columbia late in February, and Nova Scotia and southwest Ontario in mid-March. Ten days later it is into the Montreal-Ottawa area, the Rocky Mountain trench, southern Alberta, and south-central Alaska. Thereafter, a remarkable difference appears between East and West. In the prairies, the Athabasca-Mackenzie basin, and the interior plateaus of British Columbia, the thaw travels rapidly northward. By April 25, it has passed Fairbanks in central Alaska, and by May 10 has reached the eastern shores of Great Slave Lake and Great Bear Lake. In the east, prevailing air flow from the northwest, combined with the mass of ice on Hudson Bay, retard the progress of spring, and by May 10 thawing temperatures have reached only central Labrador-Ungava and the Hudson Bay coast of Ontario.

The advance of the above-freezing conditions slows near the Arctic tree line west of Hudson Bay. Due to perennial ice in the Arctic Ocean, the Arctic coast of Alaska does not attain a mean of 32°F (0°C) until about June 10. By June 15, all but the highly glacierized plateau of the Arctic Islands experience thawing temperatures. By the end of June, all parts of the Arctic lands enjoy above-freezing mean temperatures.

Summer: The frost-free period of high summer, when mean daily temperatures are above 32°F (0°C), is short. In some areas it lasts fewer than 50 days, and through the forested regions is usually 60–100 days. In small agricultural lands, such as the Matanuska Valley just north of Anchorage, Alaska, it is about 90–120 days. On the cool, cloudy Pacific coast, much longer periods, up to 240 days, are frost free. In the East, the agricultural lands of the Montreal plain, southernmost Ontario, and the Maritime Provinces have equally long frost-free periods.

CLOUDS

Probably the easiest and most enjoyable way to become a weather watcher is simply to observe the clouds. They are, in effect, a weather station aloft, revealing what is going on at different levels of the atmosphere, and they give indications of what may happen in the hours and days to come. Low-level cumulus clouds, middle-altitude altostratus clouds, and high-flying cirrus clouds all convey information about their respective atmospheric realms. "It is by observing the changes and transitions of cloud form that the weather may be predicted," advised Luke Howard, a founding father of British meteorology. As a bonus, clouds are often spectacularly beautiful in form and color, making them a delight to study.

The History of Cloud Study: Several ancient writers recognized the variety of cloud shapes in the sky as well as their usefulness in predicting weather. Early in the fourth century B.C., the Greek philosopher and naturalist Theophrastus of Eresus (on Lesbos) made such general observations as: "If in fair weather a thin cloud appears stretched in length and feathery, the winter will not end yet," and, "Fear not as much a cloud from the land as from ocean in winter; but in

the summer a cloud from a darkling
coast is a warning."

Despite such acute observations, no one
attempted to classify the multiform
clouds in a scientific way until about
1800, when, almost simultaneously, a
Frenchman and an Englishman
proposed systems of nomenclature. In
France, the naturalist Chevalier de
Lamarck (1744–1829) published a list
of cloud types in 1802; his product,
however, was flawed by the inclusion of
references to the moon and its supposed
influence on the weather, which did not
find acceptance among his colleagues.
Then, in 1803, Luke Howard (1772–
1864), an English chemist and weather
observer, submitted an essay to his local
literary club entitled "On the
Modification of Clouds." Howard's
work quickly won recognition for its
logical arrangement and use of
internationally understood Latin
designations. The most enthusiastic
advocate of Howard's system was the
German poet Goethe, who in 1817
wrote "Wolkengestalt nach Howard"
(Cloud Forms According to Howard).
He also dedicated four nature poems to
the English student of clouds.

Howard's three main cloud
"modifications," or classifications, as
we would call them today, were *cirrus,*
for parallel, threadlike clouds, either
converging or diverging; *cumulus,* for
protuberant or domelike masses that
grow upward from a horizontal base;
and *stratus,* for an extensive cloud cover
that is continuous and grows from top
to bottom. Howard assumed that
clouds could change from one type into
another, giving rise to combinations
such as cirrostratus, cirrocumulus, and
stratocumulus. He added as a fourth
type a rain cloud, calling it *nimbus*—
hence our term cumulonimbus, for a
thunderhead. Howard's basic
classifications have been followed
for nearly two centuries, with

various minor rearrangements by other observers.

The first comprehensive collection of cloud photographs and descriptions, called the *International Cloud Atlas,* resulted from a meeting of representatives of various national weather services in 1896. Sixty years later, in 1956, the Geneva-based World Meteorological Organization published a two-volume revision of the original work, with color photographs, that incorporated a greatly expanded knowledge of clouds gained through aviation experience. The atlas was revised again in 1987, and remains the standard reference for meteorological services the world over in identifying cloud conditions for the daily weather map.

How Clouds Form: Clouds form when moist air is cooled. This can occur in a number of ways. Often, cooling is due to convection, whereby unequal heating of the ground surface creates rising air currents. As it ascends, the air expands and cools. Eventually it reaches its *dew point,* or the temperature at which the invisible water vapor contained in the air condenses into a collection of tiny water droplets. We see these water droplets from the ground as a cloud. If the droplets continue to acquire moisture and grow large enough, they fall from the cloud as rain.

Another way clouds are formed is through the "lifting" of air by a storm system or, more often, by the slope of a hill or a mountain, in which case the process is known as *orographic lifting.* In rising up the slope, the air current cools; if it reaches its dew point or condensation level, a fog forms. If this fog is above the surface of the earth, it forms a cloud. Easterly wind currents blowing from the western Great Plains, for example, often form extensive cloud sheets as they move toward the lower

Cloud Formation

Clouds Formed by Convection

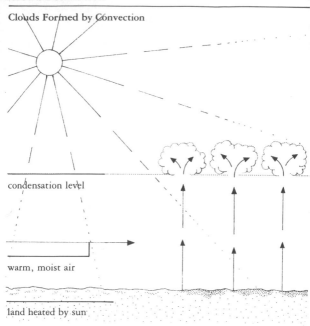

condensation level

warm, moist air

land heated by sun

Clouds Formed by Orographic Lifting

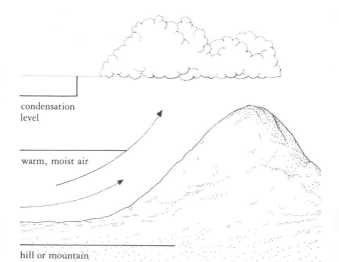

condensation level

warm, moist air

hill or mountain

slopes of the Rocky Mountains. Whether the air is stable or unstable is all-important in cloud formation. Air becomes unstable when it is heated from below by passing over land or water that has been warmed by solar radiation. The heated air rises as described above, clouds form, and they are carried downwind. Stable air develops when the atmosphere becomes stratified, with a layer of warm air overlying a layer of cold air. Along the meeting line, a thin cloud will form; it may thicken into a solid layer of the stratus type. This arrangement creates an inversion, so called because it is the opposite of the more typical situation, in which cold air lies above warm air. In an inversion, the warm air puts a "lid" on the lower atmosphere, and convection cannot take place. If the layered air at the earth's surface is cooled enough overnight, fog may occur. The fog, which is actually a cloud on the ground, will persist until the rising sun warms the air to above the condensation point and the fog evaporates.

General Cloud Groups: For forecasting purposes, cloud students have divided Luke Howard's cloud classifications into two groups: heaped clouds, resulting from rising unstable air currents; and layered clouds, resulting from stable air. Clouds are also classified according to their height above the ground (high, middle, or low altitudes), and by whether or not they produce precipitation.

Cumulus, or Heaped, Clouds: Heaped clouds, known as cumulus, are flat-based with a cauliflower-like dome created by ordinary convection. The base of a cumulus cloud forms at the condensation level of the rising warm current. If the upward current continues, the dome will develop turrets, and then the cloud is called a cumulus congestus, or "towering cumulus." It may produce light showers. If the convective process continues energetically, a large cumulonimbus may result, with the possibility of a thundershower or thunderstorm, including heavy rain, hail, and lightning. As long as the convection continues, the cumulonimbus will develop actively in size and intensity for about 30 minutes to an hour. Thunderstorms sometimes form in lines several hundred miles in length along an advancing cold front.

Stratus, or Layered, Clouds: Layered clouds appear as wide sheets, or strata, with minimal vertical and extended horizontal dimensions. Sometimes they cover the entire sky, to the horizon and beyond, like a formless blanket. There is little or no convection present. They are created when layers of air of different temperatures come into close contact with each other. This can be caused by the lifting of the air layer in a cyclonic storm system, or by the rising terrain, which reduces the temperature of the air to its condensation level.

Heaped and layered clouds sometimes coexist. Low-level stratiform clouds often develop small convective cells, which produce alternately thick and thin areas in the cloud sheet. The resulting clouds are called stratocumulus, and their tops are often arranged in rows across the sky. At middle altitudes, these are called altocumulus. At much higher altitudes, above the freezing level, the same pattern can form in a cloud composed of supercooled water and ice crystals, creating a cirrocumulus cloud. The individual convective cells in cirrocumulus and altocumulus clouds appear to be much smaller than those of stratocumulus because of their distance from the observer.

Nimbus, or Rain, Clouds: Nimbus is a broad category of layered low- to middle-altitude clouds, from which precipitation is falling. Nimbus clouds are generally dark and amorphous in appearance and have ragged edges. Small cloud fragments, called scuds, often develop beneath the main nimbus cloud from the re-condensation of evaporated moisture from raindrops; they race along in a menacing manner. On mountainsides, nimbus clouds often can be seen ascending or descending the slopes as a rainstorm arrives or departs.

High-Level Clouds: Clouds are further classified according to their usual altitude and are distinguished as high-, middle-, and low-altitude clouds. These altitudes apply to the temperate zone; the defining altitudes are lower in polar regions and higher in the tropics. The abbreviations of cloud genera in parentheses are used on daily weather maps.

High clouds generally have bases above 18,000 feet (5,486 m). They are white and thin, can be separated or detached, form delicate veil-like patches or extended wispy fibers, and often have a feathery appearance. They exist in environments well below freezing and are composed of ice crystals. The three principal types of high-altitude clouds are cirrus, cirrocumulus, and cirrostratus.

Cirrus (Ci): Detached clouds at elevations of 18,000 feet (5,486 m) and higher, taking the form of delicate white filaments, strands, or hooks, are called cirrus. They usually have a fibrous appearance. Sometimes, bands of cirrus clouds seem to emerge from a single point on the western horizon and spread across the entire sky.

Cirrus clouds are composed almost exclusively of ice crystals. Their fibrous appearance results from the wind "stretching" streamers of falling ice particles. The feathery strands of cirrus, called mares' tails, often warn of the

Cloud Types

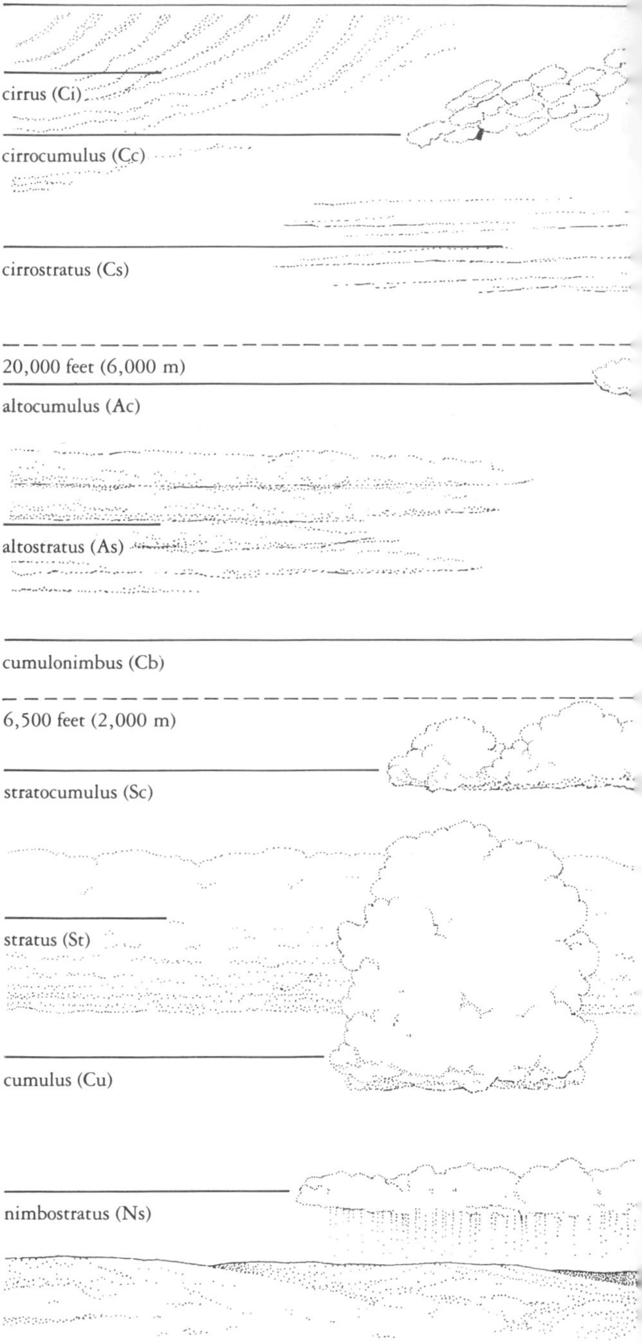

cirrus (Ci)

cirrocumulus (Cc)

cirrostratus (Cs)

20,000 feet (6,000 m)

altocumulus (Ac)

altostratus (As)

cumulonimbus (Cb)

6,500 feet (2,000 m)

stratocumulus (Sc)

stratus (St)

cumulus (Cu)

nimbostratus (Ns)

approach of a warm front marking the advance of a storm system. Snow crystals may fall from thicker, darker cirrus clouds, but they usually evaporate in the drier air below the cloud. The precipitation trail left by the crystals is known as virga. Cirrus clouds can be seen at close hand from the window of a jet plane flying above 25,000 feet (7,620 m).

Cirrocumulus (Cc): Clouds appearing in thin white patches, sheets, or layers without shading are called cirrocumulus. They are composed of very small tufts, grains, or ripples that may be merged or separate and are arranged in a roughly wavelike or dappled structure. Most of the cloud elements have an apparent width of less than one degree of arc. (One degree of arc is twice the apparent size of the full moon. For comparison, 90 degrees of arc extends from the horizon to zenith.) Cirrocumulus clouds are transient formations, often developing from cirrus clouds or cirrostratus clouds (see below) and soon returning to their original forms. To some observers, their wavelike pattern resembles the scales of a mackerel, hence the popular term "mackerel sky."

Cirrostratus (Cs): These clouds form a translucent, whitish veil of either fibrous or smooth appearance that totally or partially covers the sky, often developing into a featureless sheet that stretches across its entire breadth. Cirrostratus clouds are capable of producing haloes around the sun or moon while not entirely obscuring their light. The cloud sheet may develop to a thickness of 2,000 to 3,000 feet (610–914 m) and blend into the clouds of an approaching warm front. If the sheet merges into an area of altostratus clouds (see below), precipitation may be in the offing.

Middle-Level Clouds: Between altitudes of 6,000 and 18,000 feet (1,829–5,486 m) are clouds that form the main portion of active weather

systems moving across the country.
Middle-level clouds may resemble
higher-altitude cloud types in general
appearance, but are composed mostly of
liquid water droplets. Middle-altitude
clouds are generally of the stratus or
layer type, indicating that the air at
these levels is stable and without
vertical currents. They usually have
the prefix "alto-," in meteorology
meaning middle. The principal types of
middle-level clouds are altocumulus
and altostratus.

Altocumulus (Ac): Middle-level clouds that form in white
or gray patches, sheets, or layers and
are composed of rounded elements,
generally with shading, are called
altocumulus. Often they occur in a
wavelike arrangement, sometimes
resembling a honeycomb. Occasionally,
they are partly fibrous or diffuse and
may or may not be merged; most of the
regularly arranged small elements
usually appear to have a width of
between one and five degrees of arc.
Altocumulus clouds are almost
invariably composed of water droplets,
but they very seldom produce
precipitation. When they pass over
rough or mountainous terrain, the air
flow may be forced to undulate up and
down, resulting in the formation of a
species of altocumulus known as a
lenticular, or lens-shaped, cloud. These
have been described as resembling a
flying saucer.

Altostratus (As): These clouds form in a uniform grayish
or bluish sheet having very little
texture. They are usually thicker,
grayer, and lower than cirrostratus. The
cloud sheet washes out the sky, giving
rise to the term "watery sky." The
sheet may cover the sky partially or
totally; often parts are thin enough to
reveal the sun, at least vaguely, so that
it appears as it would through ground
glass. Altostratus clouds are generally
1,000 to 3,000 feet (about 300–900 m)

thick and do not produce halo phenomena. Sometimes, they acquire excess moisture from cooling and transform into nimbostratus, the rain cloud, which can produce varying amounts of precipitation.

Nimbostratus (Ns): A gray, often dark, cloud layer whose appearance is usually rendered diffuse by falling rain or snow, nimbostratus usually covers the entire sky and is thick enough to blot out the sun. Ragged clouds or shreds of clouds, called scuds, frequently occur below the main layer, indicating imminent wind and precipitation. The rain or snow produced by nimbostratus is steady and persistent, unlike the showers produced by some cumulus clouds. Nimbostratus is often classified as a low-altitude cloud, but it can appear at middle altitudes during a rainstorm.

Low-Level Clouds: Low-altitude clouds are found below 6,000 feet (1,829 m). They are either flat, layered stratus, or cumulus with flat bases and rounded tops. These cumulus clouds can transform into middle- or even high-level clouds when lifted by convective currents.

Stratocumulus (Sc): Clouds that form in low, distinct, gray or whitish patches are termed stratocumulus. The patches have a well-defined, rounded appearance and are often merged or organized into rounded masses, rolls, or regular long sheets. Sometimes, stratocumulus clouds are formed from the spreading out of cumulus. Their flat, even bases have dark patches and tessellations (checkerboard patterns), while their tops are rounded. Most of the regularly arranged small elements have an apparent width of more than five degrees of arc.
The persistence of stratocumulus is important in regulating the temperature of the lower atmosphere, especially over vast expanses of

subtropical oceans, such as the Pacific off the California coast, where these types of clouds are common. Stratocumulus cover prevents a significant amount of solar radiation from reaching the surface of the ground by day and greatly diminishes the escape of terrestrial heat by night. Frequently, these clouds are arranged in bands, or "rolls," lying across the wind, thus indicating the wind's direction at cloud height.

Stratus (St): Stratus clouds at low altitudes generally create a flat layer with a fairly uniform base. The limited depth of the cover inhibits the production of precipitation in any quantity, but light forms, such as drizzle, ice prisms, or snow grains, may occur. Seen on nearby hills, the cloud appears to be fog. The sun's outline may be visible through stratus clouds.

Cumulus (Cu): Fair-weather, or cumulus, clouds are detached from one another and generally have well-defined, flat bases and domed tops resembling cauliflower. Their outlines are sharp, and they often develop vertically in the form of rising mounds, domes, or towers. The sunlit parts are brilliant white; the base is relatively dark and roughly horizontal. Under normal conditions, the bases of cumulus clouds form at the same altitude, which is the condensation level of the rising air currents, so that, when viewed at a distance, their flat bottoms seem to merge into a level plane.

Cumulonimbus (Cb): Forming heavy, dense columns, cumulonimbus clouds rise from low altitudes to considerable heights. They grow from a swelling mass, known as cumulus congestus, within which there is great turbulence caused by rising and falling currents. The cumulus congestus grows into a mountainous mass, often exhibiting multiple turrets. Upon reaching the upper inversion at the tropopause level, the topmost portion spreads out in an anvil shape; downwind in the fast upper airflow, it forms a long plume of cloud that is visible at great distances. Dark bases and these white or gray anvil tops are the trademarks of the cumulonimbus and a sure sign of severe weather to come. Within the cloud, raindrops caught in ascending air currents may freeze, producing hailstones of increasingly large size. Some of these are tossed out and fall to earth; others become so heavy that the rising currents cannot sustain their weight, whereupon they, too, fall to earth. Under the base of this often dark, menacing cloud, there are frequently low, ragged clouds, blown in seemingly contrary directions by gusty, shifting winds. Precipitation, in the form of rain or hail, can be heavy.

Cloud Formation Definitions: Below are some definitions of cloud species, varieties, supplementary features, and accessory clouds. The first fourteen definitions, from *fibratus* through *capillatus,* are of cloud species, grouped together by peculiarities in shape and internal structure. The next nine definitions, from *intortus* through *opacus,* are of cloud varieties, grouped by their characteristics of organization and visibility. The final nine definitions, from *incus* through *pannus,* are a grouping of attached or related cloud forms called supplementary features and accessory clouds.

Fibratus: Thin or detached clouds consisting of nearly straight or irregularly curved filaments. Applies mainly to cirrus and cirrostratus.

Uncinus: Cirrus clouds often shaped like a comma that terminate at the top in a hook or in a tuft.

Spissatus: Cirrus clouds that appear grayish because of their thickness.

Castellanus: Clouds that present, in at least some portion of their upper part, cumuliform protuberances in the form of turrets. The turrets are connected by a common base and seem to be arranged in lines. Applies to cirrus, cirrocumulus, altocumulus, and stratocumulus.

Floccus: A cloud in which each unit is a small tuft with a cumuliform appearance, with a ragged lower part often accompanied by virga. Applies to cirrus, cirrocumulus, altocumulus, and sometimes also in stratocumulus.

Stratiformis: A horizontal sheet of clouds. Applies to altocumulus, stratocumulus, and, occasionally, to cirrocumulus.

Nebulosus: A thin layer of clouds without distinctive characteristics. Applies mainly to cirrostratus and stratus.

Lenticularis: Clouds shaped like lenses or almonds that are occasionally iridescent. Lenticularis clouds are usually orographic; however, they may also occur in regions with level terrain. Applies mainly to cirrocumulus, altocumulus, and stratocumulus.

Fractus: Clouds that appear as irregular shreds. Applies only to stratus and cumulus.

Humilis: Slightly vertical cumulus clouds that generally appear flattened.

Mediocris: Moderately vertical cumulus clouds whose tops show fairly small protuberances.

Congestus: Extremely vertical cumulus clouds whose bulging upper part often resembles a cauliflower.

Calvus: Cumulonimbus clouds in which some protuberances on the upper part are beginning to lose their cumuliform outlines, but in which no cirriform parts can be distinguished. The protuberances tend to form a whitish mass with somewhat vertical striations.

Capillatus: Cumulonimbus clouds characterized by the presence of distinctly fibrous or striated cirriform, frequently having the form of an anvil or a plume. Cumulonimbus capillatus clouds are usually accompanied by a shower or thunderstorm, often by squalls, and sometimes by hail. They frequently produce virga.

Intortus: Cirrus clouds whose filaments are irregularly curved and entangled.

Vertebratus: Clouds arranged in a manner suggestive of vertebrae, ribs, or the skeleton of a fish. Applies mainly to cirrus.

Undulatus: A cloud composed of separate or merged elements and organized in undulations. Applies mainly to cirrocumulus, cirrostratus, altocumulus, altostratus, stratocumulus, and stratus.

Radiatus: Clouds arranged in parallel bands that seem to converge toward a point on the horizon or, when the bands extend across the sky, toward two opposite points on the horizon called "radiation points." Applies mainly to cirrus, altocumulus, altostratus, stratocumulus, and cumulus.

Lacunosus: Thin cloud patches, sheets, or layers marked by round holes with fringed edges. Applies mainly to cirrocumulus and altocumulus; may also apply, although very rarely, to stratocumulus.

Duplicatus: Coinciding cloud patches, sheets, or layers, sometimes partly merged at slightly different levels. Applies mainly to cirrus, cirrostratus, altocumulus, altostratus, and stratocumulus.

Translucidus: A translucent patch, sheet, or layer of clouds through which the sun or moon can be seen. Applies to altocumulus, altostratus, stratocumulus, and stratus.

Perlucidus: An extensive patch, sheet, or layer of clouds with well-defined but sometimes small spaces between its elements. The sun, moon, sky, or overlying clouds can be seen through the spaces. Applies to altocumulus and stratocumulus.

Opacus: An extensive patch, sheet, or layer of clouds whose greater part is sufficiently opaque to completely obscure the sun or moon. Applies to altocumulus, altostratus, stratocumulus, and stratus.

Incus: A cumulonimbus cloud's upper portion organized as an anvil with either a smooth, fibrous, or striated appearance. It is a supplementary feature.

Mamma: Hanging protuberances that look like round pouches on the surface of a cloud. These supplementary features occur mostly with cirrus, cirrocumulus, altocumulus, altostratus, stratocumulus, and cumulonimbus.

Virga: Streaks of rain or ice crystals that fall from a cloud but evaporate before reaching the ground. Applies to altocumulus, altostratus, and high-level cumuliform.

Praecipitatio: Generating precipitation that falls from a cloud and reaches the ground. This occurs with altostratus, nimbostratus, stratocumulus, stratus, cumulus, and cumulonimbus.

Arcus: A thick, horizontal accessory cloud with ragged edges, located on the lower front part of the main cloud. Occurs with cumulonimbus and cumulus.

Tuba: A cloud column that hangs from a cloud base and, upon reaching the earth's surface, forms a tornado or waterspout. This applies to cumulus and cumulonimbus.

Pileus: A small horizontal accessory cloud, in the form of a hood, that occurs above or on top of a cumuliform cloud. Occurs with cumulus and cumulonimbus.

Velum: A thin layer of accessory clouds pierced by cumuliform clouds. Occur with cumulus and cumulonimbus.

Pannus: A layer of shredded accessory clouds below the main cloud. Occurs with nimbostratus, cumulus, and cumulonimbus.

STORMS

A storm is any kind of disturbance in the weather—and especially one that brings with it unpleasant, or even violent, atmospheric conditions. Meteorologists recognize many different kinds of storms, which vary not only in their causes but also in size, intensity, and duration.

In synoptic meteorology—the kind of moment-by-moment observation that provides "snapshots" of the weather at a particular place and time—storms tend to be regarded as integral complexes of weather conditions. Thus, a thunderstorm, for example, may be defined as a distinct, individual event made up of a network of wind, precipitation, clouds, pressure, and other key influences or ingredients. Broken down into its component parts or conditions, the thunderstorm can be analyzed in such a way that meteorologists are able to predict, with fair accuracy, not only the effect such a storm will have but also the likelihood of its occurring again, given the proper conditions.

In everyday terms, however, we think of a storm as a transitory event, distinguished less by its meteorological conditions than by its drama or violence. In this way, we call a very heavy downpour, or a spectacular blizzard, a storm.

In the essays that follow, you will find information about the principal types of storms that occur in North America. These vary from the major disturbances whose effects are often felt throughout a wide segment of the continent— thunderstorms, tropical depressions, and hurricanes—to such intensely local events as hailstorms, wind storms, and dust storms.

THUNDERSTORMS

Thunderstorms have been called "weather factories" because they can produce a great variety of weather phenomena—showery rain, pelting hail, strong and gusty winds, sudden changes of temperature at the ground, shafts of lightning, peals of thunder, and even tornadoes. In a few minutes' time on a windless, hot summer afternoon, a thunderstorm can transform the local meteorological scene from still and oppressively close to windy and comfortably cool.

The essential ingredient in the creation of a thunderstorm is a warm, moist, unstable atmosphere. Thunderstorms form when an air parcel becomes buoyant, rises, and continues to rise as long as it remains warmer than the surrounding air. There are several different ways in which this instability can come about.

Some storms, known as *air mass thunderstorms* or *convective thunderstorms,* are created entirely by local conditions, in which surface heating causes the required instability and buoyancy. Another type, called a frontal thunderstorm, results from the turbulence attending the arrival of a cold front in the area. The underrunning cold air of the front causes the warm air at the surface to

Development of a Thunderstorm

Cumulus Congestus Stage

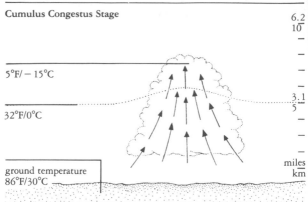

5°F/ − 15°C

32°F/0°C

ground temperature
86°F/30°C

6.2
10

3.1
5

miles
km

Mature Cumulonimbus Stage

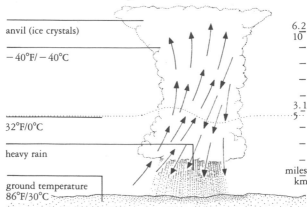

anvil (ice crystals)

− 40°F/ − 40°C

32°F/0°C

heavy rain

ground temperature
86°F/30°C

6.2
10

3.1
5

miles
km

Dissipating Stage

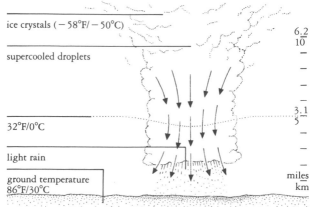

ice crystals (− 58°F/ − 50°C)

supercooled droplets

32°F/0°C

light rain

ground temperature
86°F/30°C

6.2
10

3.1
5

miles
km

become unstable and rise in a column. Frontal thunderstorms are often severe, forming in long lines and affecting a wide territory as the front advances.

Development: The rising column of warm air described above develops into what is called a thunderstorm cell, an entity full of updrafts that constitutes the core of the thunderstorm. Such a cell has three stages in its life cycle (see the illustration on page 96). In the first, the cumulus stage, the cell consists of upward-moving air currents that extend from below the cloud base to the cloud top. If the air is sufficiently unstable, the upward currents will continue to rise, drawing in warm, moist air and causing water vapor to condense into raindrops. The release of latent heat that occurs during this process supplies the energy required to continue the updrafts, increase their intensity, and expand the size of the thunderstorm cell.

With an increase in size comes an increase in the water content of the cloud and a growth of the size of the raindrops or ice crystals forming in the updraft. When, through accretion, these drops or crystals become heavy enough, they overcome the support supplied by the updraft and fall, forming a downward current alongside the updraft. At this point, the cell has reached the mature stage, with precipitation descending and fresh moisture ascending. Meanwhile, the top of the updraft may reach the tropopause, at an altitude of 7–11 miles (12–17 km), and spread out to form an anvil cloud, the trademark of a fully developed cumulonimbus.

Eventually the cold, descending column of air and precipitation expands outward near the ground and shuts off the updrafts. (Sometimes the downdraft takes the form of a violent downburst of shifting wind, which can be hazardous to landing or departing aircraft.) After a life span of up to two hours, the thunderstorm cloud dissipates and, generally, the local atmosphere clears.

Season: Thunderstorms occur most frequently during the warm months of the year, except along the Pacific Coast, where they form in connection with warm and cold fronts in winter storms. Regions east of the Mississippi River experience the greatest number of storms in July, when the combination of surface heating and atmospheric moisture is at its peak. Over much of the Great Plains region, June is the month of maximum thunderstorm frequency. Later, the moisture source from the Gulf of Mexico is shut off by the summertime westward extension of the Atlantic anticyclone, causing storms in the southeastern states. The northern Rocky Mountains and the intermountain region experience maximum thunderstorm activity in July; the southern Rockies favor August, when Gulf moisture arrives on the wings of a southeasterly airflow around the summer anticyclone. Frontal thunderstorms can occur in the northern states at any time of the year when strong cyclonic activity brings tropical air northward, although such storms are usually limited to flashes of cloud lightning and brief thunder rolls.

Areas of   The frequency with which
Occurrence: thunderstorms occur varies from one region to another. The area with the greatest average number of thunderstorms each year is the interior of the Florida peninsula, where the Atlantic and Gulf airstreams converge; Lakeland, Florida, for instance, located in the center of the state, averages 100 thunderstorm days per year. Thunderstorms occur on more than 70 days a year along the Gulf Coast from Louisiana east to Florida and south

along its peninsula almost to the tip. To the northwest, over the middle Mississippi Valley and east-central Great Plains, there are 50 or more thunderstorm days annually. Most of the Northeast experiences fewer than 30 days of thunderstorms per year, and eastern New England has fewer than 20 such days. The Great Lakes region generally records between 30 and 40 days of thunderstorms per year. A secondary zone of relatively high annual storm frequency (more than 50 storm days), caused by high mountain topography, lies over the central Rocky Mountains in Colorado and New Mexico. Toward the west, the frequency decreases steadily, to fewer than 10 thunderstorm days in the Pacific states and fewer than 5 storm days along the coast, where cool Pacific airstreams inhibit convection.

Time of Occurrence: Most local air-mass thunderstorms occur in the late afternoon and evening hours, when the destabilizing effect of the heated earth is at its maximum. Frontal thunderstorms may occur at any time of day or night, although they, too, are most severe in late afternoon or early evening. Thunderstorms forming over the Rocky Mountains in daytime may drift eastward over the Great Plains during the evening and night and still remain severe after midnight; such storms are sustained by radiational cooling from the top of the cloud into the dry atmosphere above, an effect that maintains instability within the cloud. In most areas in the east, local thunderstorm activity usually decreases in intensity and frequency after sundown.

Lightning: If thunder is the atmosphere's noisiest production, lightning is its most dazzling. The two phenomena have a direct cause-and-effect relationship. Lightning has been described as an

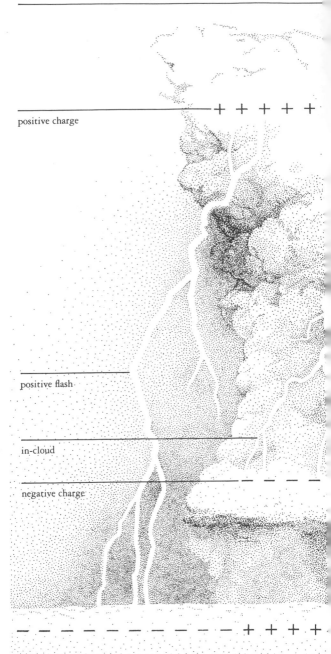

positive charge

+ + + + +

positive flash

in-cloud

negative charge

- - - - - - - - - - + + + +

cloud-to-cloud

cloud-to-ground

enormous electric spark between regions of oppositely charged particles. It develops within a growing cumulonimbus cloud as negatively charged electrons migrate from freezing ice crystals to liquid water droplets, resulting in a predominantly negative charge in the lower portions of the cloud and positive charges in the icy upper portions. Because like charges repel each other, the negative charge in the lower cloud pushes electrons on the ground outward to the periphery of the storm, leaving a residual positive charge directly beneath the cloud. These are often called "positive flashes."

Electrical potentials, or voltages, build up between all regions of opposite charge. When the potentials become large enough, electrical discharges occur between the oppositely charged regions. (See the illustration showing the different kinds of lightning on pages 100–101.) Most lightning activity takes place within the cloud itself and is thus called *in-cloud lightning;* only about 20 percent of discharges are the cloud-to-ground variety.

The familiar *cloud-to-ground lightning* begins with a negative discharge from the cloud, known as a *leader.* When a leader approaches the ground, an upward flow of positive electrical charge, known as the *return stroke,* rushes to meet it. There may be several return strokes, but to our eyes they appear as a single stroke, or what is called *streak lightning.* If the light channel is branched, it is known as *forked lightning;* when the flash occurs within the clouds and the strokes are obscured by intervening clouds, we call it *sheet lightning* or *cloud-to-cloud lightning.* If the flash is so distant that we see it light up the sky only above the storm, and we cannot hear the thunder, it is called *heat lightning. Ball lightning,* a rare phenomenon, has been reported by observers as a luminous sphere, about 6 inches in diameter, that appears and persists for a few seconds after a close cloud-to-ground discharge. The ball may roll along and commit a variety of pranks before it fizzles out; rarely, it explodes. There is no satisfactory scientific explanation for ball lightning, but when it is understood, the knowledge of how electricity contains itself in a ball for several seconds may be helpful in efforts to develop new means of energy production and transmission.

Thunder: Much of the energy of a lightning discharge is expended in heating the atmospheric gases in and immediately around the luminous channel. In just a few microseconds, the temperature of the gases rises to about 18,000°F (about 10,000°C), causing the air to expand violently. The resulting pressure waves are followed by a succession of compressions induced by the inherent elasticity of the air; we hear them as thunder—*thunor* in Old English, akin to the Latin verb *tonare,* to thunder.

Thunder is seldom audible at points farther than 15 miles (about 24 km) from a lightning discharge; 25 miles (about 40 km) is approximately the upper limit, and 10 miles (about 16 km) is a fairly typical range of audibility. At such distances, thunder has a characteristic low-pitched rumbling sound. The pitch is due to strong absorption and scattering of the high-frequency components of the original sound waves, while the rumbling results from the fact that the sound waves are emitted by different portions of the lightning channel, which lie at varying distances from the listener.

Because light travels so fast (approximately 186,000 miles per second), you can see a lightning flash

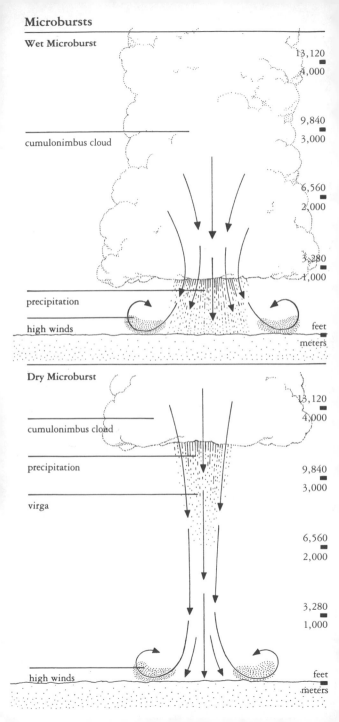

Microbursts

Wet Microburst

13,120
4,000

9,840
3,000

cumulonimbus cloud

6,560
2,000

3,280
1,000

precipitation

high winds

feet
meters

Dry Microburst

13,120
4,000

cumulonimbus cloud

precipitation

9,840
3,000

virga

6,560
2,000

3,280
1,000

high winds

feet
meters

almost the instant it happens. Sound, however, is slower; sound waves normally travel about 1,090 feet per second, so it takes about 5 seconds for thunder to travel 1 mile (about 1.6 km). You can estimate the number of miles between you and a thunderstorm by counting the number of seconds between the flash of lightning and the arrival of the sound of thunder, and then dividing by five.

Microbursts: Downdrafts have long been recognized as contributing to the dynamics of thunderstorms and convective showers. Recent research has associated them with several aircraft accidents that have occurred during takeoff or landing. The structure of these microbursts, as they are now called, and the conditions that create them are the focus of much current study.

A microburst is a downdraft that is less than 2.5 miles (4 km) wide and that causes sudden changes in wind speed and direction, called *wind shear.* Pilots of departing or landing aircraft must avoid these shifts or risk fatal results. Microbursts also pose hazards to small sailboats, which can be capsized by strong, suddenly shifting winds; these shifts also have been known to fan forest fires in unexpected directions, imperiling the lives of firefighters.

Dry Microbursts: A great variety of conditions can produce microbursts, but two extreme situations seem to create them in large numbers. One is a very dry environment, as is common over the semiarid western Great Plains and intermountain region, where moist convection is just barely possible. Cumulus clouds with very high bases form, having very dry air below and a moist layer above. Precipitation virtually evaporates before reaching the earth's surface (such precipitation is known as *virga*), but the evaporative cooling of the air intensifies the downdraft, producing destructive winds.

Wet Microbursts: The other environment that tends to produce microbursts is an extremely wet one in which rain-laden downdrafts must draw in drier air from outside the storm to undergo evaporative cooling. This action can produce a localized microburst that becomes embedded in a larger area of heavy rain. The necessary wet environment often attends severe weather in the central and eastern portions of the United States.

The History of Microburst Study: The microburst was first identified by Professor T. Theodore Fujita of the University of Chicago in 1974. His studies of several cases led to a full-scale research project by the National Center for Atmospheric Research (NCAR) at Boulder, Colorado. The Boulder study and others have shown that microbursts can be detected by Doppler radar systems, which have the ability to measure wind speeds in cloud formations. In the 1980s, the Federal Aviation Administration, along with the National Oceanic and Atmospheric Administration (NOAA) and NCAR, developed Doppler radar systems for installation at airports, which are expected to become operational across the United States during the 1990s. Also, an educational program aimed at incorporating a knowledge of microburst behavior into all training programs for commercial pilots was instituted in the 1980s in the interest of greatly reducing the frequency of airline accidents attributable to microbursts. During the 1970s and 1980s, there were 27 such accidents reported in the United States, involving 491 deaths and 206 injuries.

TORNADOES

If hurricanes are the largest storms on earth, tornadoes are the most violent. Nothing produces more destructive power in a restricted area than a tornado as it passes by, sweeping the ground clear of all movable objects. A tornado's devastating wind blasts put all human life in jeopardy, sending dangerous debris flying and lifting buildings from their foundations. It is small wonder the sight of these intense storms strikes terror.

Technically, a tornado (from the Spanish word *tornado,* past participle of *tornar,* meaning to turn; and from *tornada,* meaning thunderstorm) is a violently rotating column of air, usually with a small diameter, extending from a turbulent cloud to the ground. Some smaller or less violent storms also have tornadic characteristics.

A *funnel cloud* is an incipient tornado whose column does not reach the ground. A *whirlwind* is a small, local rotating column of air that does little damage but lifts up leaves and paper as it briefly dances along a street or across a field. (Whirlwinds are popularly known as twisters, a name used by the public for small tornadoes but frowned on by meteorologists.)

Dust devils and *sand devils,* familiar sights in dry, desert country, are

Structure of a Tornado

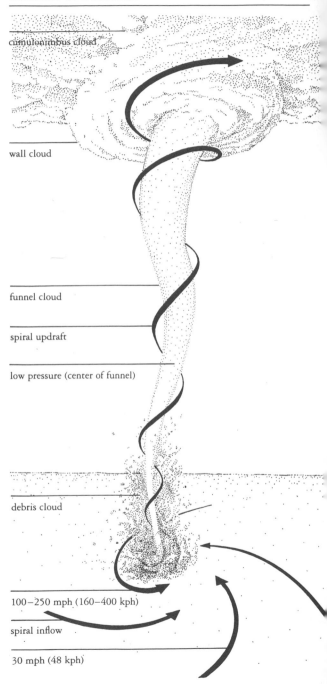

cumulonimbus cloud

wall cloud

funnel cloud

spiral updraft

low pressure (center of funnel)

debris cloud

100–250 mph (160–400 kph)

spiral inflow

30 mph (48 kph)

spiraling columns of dust- or sand-filled air, often several hundred feet high, that perform their antics for a few minutes. Whirlwinds and devils differ in their origin from true tornadoes. They are caused by intense local heating of the surface of the earth, whereas tornadoes are caused by clashing warm and cold air currents aloft that lead to atmospheric instability and severe turbulence.

Life Cycle of a Tornado:
During its life cycle, a tornado undergoes considerable changes in size, shape, and behavior. In fact, while all of the five general overlapping stages described below have been seen, each tornado may have individual deviations.

Funnel Cloud:
The rotating column of air usually develops within a cumulonimbus cloud and subsequently extends toward the ground. We see this stage as a rotating funnel cloud that descends from the cloud base and approaches the ground.

Tornado:
When the rotating column of air reaches the ground, by definition the disturbance becomes a tornado. We can see the funnel cloud reach the ground; sometimes we can also see dust and debris whirling on the ground before the funnel actually touches down. In weak tornadoes, particularly in dry climates, the ground-level dust whirl may be visible before the funnel cloud.

Mature Tornado:
During the tornado's mature stage, the funnel reaches its greatest width. It is almost always nearly vertical and most of the time is touching the ground, though skipping may occur along a lengthy path. At this time the tornado causes severe damage to whatever it encounters. In dry climates the funnel cloud may be no more than a small cone protruding from the cloud base, while the rotating column fills with dust and debris and becomes visible.

Shrinking Tornado:
During the shrinking stage, the funnel narrows and tilts away from its vertical

position. At this time the path of damage becomes smaller.

Decaying Tornado: As the tornado decays, the funnel stretches into a rope shape and the visible portion becomes contorted and finally dissipates. This stage is often called the *rope stage* because of its appearance.

Funnels: The distinguishing feature of a tornado is the funnel. This generally cone-shaped column narrows down to the ground from a parent cloud that is almost always a cumulonimbus of an active thunderstorm. The visible funnel is formed by the condensation of water vapor resulting from the lower pressure within the whirl and from debris drawn into the spinning mass as the size and strength of the funnel increase. Several funnels may develop in a mature tornado system, with small vortices continually forming and dissipating while whirling around the central core of the main tornado circulation. A tornado funnel can assume various forms, from a thin, writhing, ropelike column of grayish white to a thick, amorphous mass of menacing black. In the Northern Hemisphere, tornadoes almost always spin counterclockwise, although verified instances of clockwise circulation have been reported. Films of tornadoes have revealed that the visible flow on and just outside the funnel surface is nearly always spiraling upward.

Origins: In recent years, the exact mechanism that causes a tornado to form, called *tornadogenesis,* has been the subject of increasingly fruitful research. Nevertheless, some mystery still surrounds tornadoes, and their formation cannot be predicted with absolute accuracy, even when conditions for their occurrence seem just right. Tornadoes are usually associated with thunderstorm conditions. They require a moist airstream that is warm for the season and usually comes from a southerly direction. A 1986 survey has shown that most tornadoes occur when air temperatures are above 65°F (18°C) and the dew point (temperature at which condensation begins) is 50°F (10°C) or higher.

Sometimes a tornado is formed by a small cyclonic circulation, called a mesocyclone or tornado cyclone, embedded in a larger storm circulation. Mesocyclones are about 10 miles (about 16 km) in diameter and contain miniature cold and warm currents that interact to form a tornado funnel.

Certain atmospheric conditions are conducive to the formation of tornadoes. One such condition is a temperature structure of the atmosphere in which warm air overlies cold air (called an *inversion*). Inversions inhibit vertical currents and help maintain the concentration of low-level moisture. The early presence of the inversion also prevents deep convection from dissipating heat energy upward. At intermediate levels of the atmosphere, a warm, dry layer usually has an air movement from the southwest; at high levels, a strong westerly flow with jet stream characteristics prevails. A graphic cross section of the entire mass of air would show deep convective instability, with large drop-offs of temperature and moisture through a great depth. Dynamic lifting in rising air currents soon destroys the inversion. Strong low-level vertical wind shear and a low-level jet flow assist in tornado formation. The contraction of the tornado cyclone into a tighter circle causes increasing wind speeds, which leads to decreased air pressure inside the cyclone and more violent updrafts. This process, known as the conservation of angular momentum, can cause winds of 50 miles per hour (80 kph) to 250 miles per hour (322 kph); the process works in much the same way ice skaters performing spins increase speed by drawing their arms in tight to their bodies.

Direction of Movement:
Generally speaking, 87 percent of all tornadoes have a preferred directional movement, which is from southwest to northeast. Rarely do tornadoes move toward the west. Some tornadoes, however, have been reported to move from any quadrant, change directions abruptly, follow zigzag paths, become stationary, or perform loops or complete circles.

Speed of Movement:
Tornadoes move forward at an average speed of 35 miles per hour (56 kph), but great variations have been recorded. Several tornadoes have been clocked at 60–65 miles per hour (97–105 kph) along the ground. During a short part of its course, the great Tri-State Tornado of March 1925 moved at the astonishing rate of 73 miles per hour (117 kph).
On the other hand, some tornadoes go nowhere. Some have been observed to remain stationary for several minutes; for example, a South Dakota tornado hovered in one field for 45 minutes. One slow mover crept along at 5 miles per hour (8 kph) for half an hour, enabling people and animals to outrun it. These stationary and slow-moving tornadoes still can cause great damage.

Speed of Wind:
No wind-measuring instrument has ever survived the impact of a full tornado. Nonetheless, it is possible to estimate wind speeds within a funnel by less direct means. For example, films and videos have been used to track the debris caught in tornadic winds. Engineers used those data to survey the damage done by tornadoes and calculate the size of the force required to inflict it. In recent years

Doppler radar, which directly measures the speed of small particles in the air, has been used to measure wind speed in and near tornadoes.

Most scientists estimate the top wind speeds in the strongest tornadoes at about 280 miles per hour (451 kph), vertical updrafts in the core of a tornado at 180 miles per hour (290 kph), and radial inflow velocities (i.e., winds being drawn directly in toward the tornado's center) at speeds of 112 miles per hour (180 kph).

Path of a Tornado:
The longest verified track (a term used by meteorologists to describe a path of a tornado) of a single tornado was made on May 26, 1917, by the Mattoon (Illinois)–Charleston (Indiana) Tornado, which traveled 293 miles (471 km). In a study of the period 1916–58, only nine paths were found to be longer than 200 miles (322 km).

A survey of almost 20,000 tornadoes shows that the average width of a tornado's path is 140.8 yards (129 m); another survey indicates that 50 percent of the path widths were 100 yards (91 m) or less, and only 2 percent were 700 yards (640 m) or more. Some large tornadoes have been a mile (1.6 km) in diameter and in the end have expanded to a whirling cyclonic mass covering 5–6 miles (8–10 km) before dissipating.

In large tornadoes, the greatest violence usually occurs in an area less than one-half mile (0.8 km) wide and frequently in a tight core about 150–200 yards (137–182 m) wide. Some of the most extreme damage occurs along circular tracks left by small but intense vortices whirling around the main tornado funnel.

Life of a Tornado:
When a tornado funnel touches the ground, its average life is less than 15 minutes; thus, a tornado traveling an average of 35 miles per hour (56 kph)

Two of the Nation's Worst Tornado Outbreaks

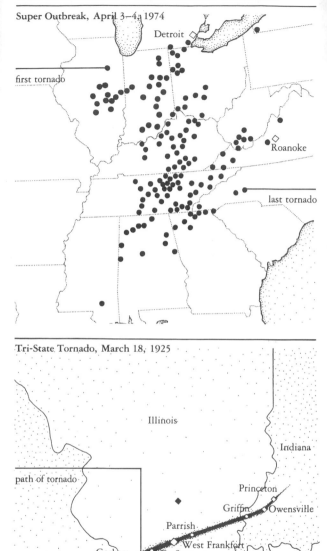

Super Outbreak, April 3–4, 1974

Detroit

first tornado

Roanoke

last tornado

Tri-State Tornado, March 18, 1925

Illinois

Indiana

path of tornado

Princeton

Griffin

Owensville

Parrish

West Frankfort

Gorham

De Soto

Murphysboro

Biehle

Redford

Annapolis

Ellington

Kentucky

Missouri

would travel a path just under 9 miles (14 kph) before dying out. The longest life recorded for a single tornado is 7 hours 20 minutes; this endurance record belongs to the Mattoon–Charleston Tornado of 1917, which also holds the distance record.

Regional Distribution:

Tornadoes occur in almost all the 48 contiguous states, although their presence is infrequent west of the Continental Divide. Funnel clouds and even tornadoes have been sighted in southern Alaska and the Canadian subarctic, as well as in Hawaii and Puerto Rico.

The area having the most frequent tornado occurrence changes with the seasons. In the winter, tornado activity is greatest through the interior of the states bordering on the Gulf of Mexico; in March and April activity advances northward into the Central Plains, the middle Mississippi Valley, and the Ohio Valley; in May and June it occurs in the northern Great Plains, upper Midwest, and Great Lakes region. The potential for activity remains there through September, when a retreat southward begins. By November and December, the region of greatest frequency is once more centered in the Gulf states.

A tornado alley of maximum activity per unit of area (typically, 10,000 square miles) extends from north-central Texas across the central sections of Oklahoma, Kansas, Nebraska, and the Dakotas; peak activity occurs between 97°W and 98°W longitudes. Florida has the greatest frequency of reported tornadoes per unit of area; however, most of the tornadoes are weak and short-lived. When we consider only the strongest, potentially most damaging tornadoes, the peak frequency of occurrence is in Oklahoma.

Canada has about one-twentieth as

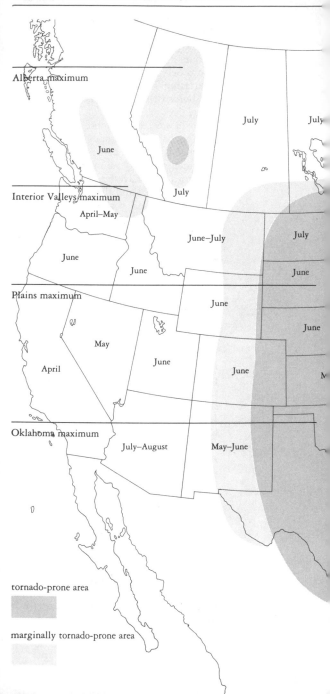

Tornado-Prone Areas and Peak Months

Alberta maximum

June

July

July

July

Interior Valleys maximum

April–May

June–July

July

June

June

June

June

June

Plains maximum

May

June

June

M

April

June

June

Oklahoma maximum

July–August

May–June

tornado-prone area

marginally tornado-prone area

June

July

August

July

Northern maximum

June–
July

July

July–
August

July–
August

June

May

June

April

April

April–July

July

May

April

April

North
Atlantic
maximum

Ohio maximum

May

April

April

July

April–May

May

April

March–
April

April

April–
May

Feb, Apr &
Nov

April &
November

June

many tornadoes as the United States.
About one-third of Canada's tornadoes
strike southern Ontario, the most
tornado-prone area being around
Windsor, Ontario. Another 40 percent
of Canadian tornadoes touch down in
the southern parts of the prairie
provinces (Alberta, Saskatchewan, and
Manitoba), at the northern extremity of
the Great Plains tornado belt. For a
breakdown of tornado-prone regions of
North America, see the map on
pages 116–117.
According to preliminary figures from
the National Weather Service, 1990
had the greatest number of U.S.
tornadoes in one calendar year, with
1,115; 1973 was second with 1,102;
1982 was third with 1,046.

Season of Occurrence: Tornadoes can develop at any time of
the year; since the year 1950 at least
one tornado has been reported on every
date of the year except January 16.
Certain times of year, however, are
more favorable for tornadoes.
January has the lowest expected average
tornado occurrence. A marked increase
takes place in March, which has twice
as many tornadoes as February, and in
April, which has two and one-half
times as many tornadoes as March. The
peak tornado month is May, when
meteorologists expect nearly 150
tornadoes; June has only slightly fewer.
On an annual basis, the seasonal
occurrence of tornadoes is clear: By
April 28, 25 percent of the expected
tornadoes normally have occurred.
June 4 marks the 50 percent point, and
by July 20 we have reached the 75
percent point.

Time of Day: Tornadoes favor the warmest part of the
day, when solar heating and
thunderstorm development are at their
maximum. Thus, 60 percent of
tornadoes occur between noon and
sunset. The median time is about

5:00 P.M. local solar time. But here, too, the tornado displays geographic variations. In the Gulf states, the hourly distribution is about evenly spread throughout the 24 hours. In the Midwest and Plains States, however, tornadoes show a strong preference for late afternoon and early evening.

Multiple Outbreaks: Tornadoes sometimes descend in swarms. The record for such a multiple outbreak during a 24-hour period was set by the Super Outbreak on April 3–4, 1974; its 148 individual tornadoes covered the area from central Alabama to southern Michigan. A total of 315 people were killed and 5,484 injured as the storms rampaged through their paths. (See the map on page 114.) On September 20–21, 1967, during Hurricane Beulah in Texas, 115 tornadoes, mostly miniature ones, were reported; five people were killed.

Most Deadly Tornadoes: The greatest tornado disaster in early America occurred at Natchez, Mississippi, on May 7, 1840, when a single large storm took the lives of 317 people, most of whom were in an area along the waterfront. According to unscientific estimates, on February 19, 1884, an outbreak of more than 60 tornadoes, spreading from the southeastern states to the Ohio Valley, killed as many as 800 people. On March 18, 1925, the great Tri-State Tornado killed 695 people in Missouri, Illinois, and Indiana. (See the map on page 114.) Furthermore, on that day seven other tornadoes struck areas in Kentucky, Tennessee, and Alabama, raising the total to 792 deaths; it was clearly the most tragic tornado day in U.S. history.

Classification of Tornadoes: Tornado researchers use a scale, known as the Fujita–Pearson Tornado Intensity Scale (named after its creators) to rate the intensity of tornadoes. The scale

Fujita-Pearson Tornado Intensity Scale

	Scale	Category	Force (mph)
U.S.	F0	Weak	0-72
	F1	Weak	73-112
	F2	Strong	113-157
	F3	Strong	158-206
	F4	Violent	207-260
	F5	Violent	261-308

	Scale	Category	Force (kph)
Metric	F0	Weak	0-116
	F1	Weak	117-180
	F2	Strong	182-253
	F3	Strong	254-332
	F4	Violent	333-418
	F5	Violent	420-496

Path Length (miles)	Path Width (yd/mi)	Expected Damage
0-1	0-17 yd	Light
1-3.1	18-55 yd	Moderate
3.2-9.9	56-175 yd	Considerable
10-31	176-556 yd	Severe
32-99	0.34-0.9 mi	Devastating
100-315	1-3.1 mi	Incredible

Path Length (kilometers)	Path Width (m/km)	Expected Damage
0-1.6	0-16 m	Light
1.6-5	16-50 m	Moderate
5.1-15.9	51-160 m	Considerable
16-50	161-508 m	Severe
51-159	0.54-1.4 km	Devastating
161-507	1.6-5 km	Incredible

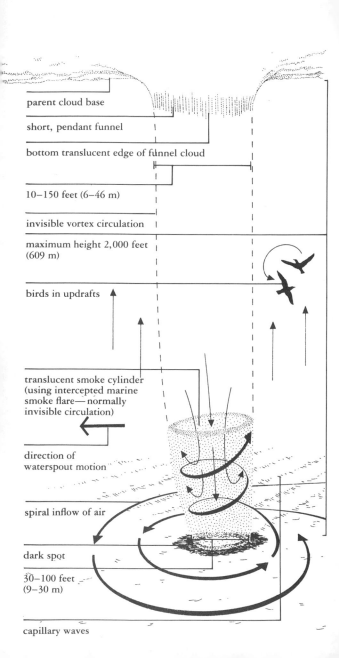

parent cloud base

short, pendant funnel

bottom translucent edge of funnel cloud

10–150 feet (6–46 m)

invisible vortex circulation

maximum height 2,000 feet
(609 m)

birds in updrafts

translucent smoke cylinder
(using intercepted marine
smoke flare— normally
invisible circulation)

direction of
waterspout motion

spiral inflow of air

dark spot

30–100 feet
(9–30 m)

capillary waves

provides ratings in each of three areas: force, or wind speed *(F)*, path length *(PL)*, and path width *(PW)*.

The force, *F,* represents the maximum wind speed within each storm. Because tornadoes destroy wind instruments, scientists use photographs of a storm's damage to calculate the force (or wind speed) that must have been present. Most news reports of tornadoes give the F value, thereby indicating the tornado's wind speed. In general, weak tornadoes are classified as *F0* and *F1,* strong tornadoes as *F2* and *F3,* and violent tornadoes as *F4* and *F5.*

The table on pages 120–121 shows the ranges of wind speed, path length, and path width based upon the Fujita–Pearson Scale. (Note: *F, PL,* and *PW* are three separate scales; for example, an *F1* classification does not mean that *PL* and *PW* are also 1.)

A survey of almost 20,000 tornadoes occurring from 1950 to 1977 for which intensity and track lengths could be determined found that 61.7 percent were weak, 36 percent were strong, and only 2.3 percent were violent. The violent storms caused 68 percent of the fatalities.

Other Whirls: In addition to tornadoes, there are several other types of weather disturbances that exhibit the typical whirling form of the tornado. By and large these other whirls are far less dangerous than tornadoes; they are nonetheless interesting, however— and some are fairly uncommon.

Waterspouts: A waterspout is a rapidly rotating column of air originating over water that exhibits tornadic characteristics. The strongest waterspouts are simply tornadoes that happen to form, or move over, a river, lake, or ocean. The large majority of waterspouts, however, differ from tornadoes in structure and life cycle. Waterspouts are usually less intense than tornadoes and normally lose their structure upon passing from water to land. Such waterspouts do not require a stormy cumulonimbus to generate, but often descend from a cumulus congestus that may or may not be precipitating. They occur most frequently in the warmer seasons and favor shallow-water locations such as the Florida Keys; however, they also have been observed on the Great Lakes, in the Hudson River, and off San Diego Bay.

Life of a Waterspout: The first visible sign of the forming vortex of a waterspout is often a dark spot or spots on the water's surface. (See the illustration on page 122.) Of the several dark spots, usually one becomes the primary spot. Smoke flares dropped on the water from the air have indicated that these dark spots, variations in the surface waves, are caused by the downward-wind rotation imposed from the air column above. In time, a short, pendant formation extending downward from the clouds may develop. This visible funnel is formed as the water vapor condenses, not as the seawater rises. (Seawater entrained in the lower part of the funnel, close to the sea's surface, is called the bush of the spout. Such entrainment is a distinct occurrence, not part of the funnel's formation.) Occasionally spiral patterns will form on the water.

The next stage begins as the wind speed increases beyond a critical value (about 50 miles per hour/80 kph). The funnel descends, throwing up a ring of spray several feet above the surface, before it increases in size, tilts, and begins moving along the surface. At the same time, the funnel tightens and increases its rotational speed. Usually, the spout height will not exceed 3,000 feet (about 914 m) and the funnel diameter will range between 50 and 150 feet (about 15–46 m). Great variations in funnel structure have been observed, and multiple spouts frequently appear close together.

The mature stage of a waterspout may last up to 15 minutes and usually is characterized by peak winds, increasing tilt, and maximum forward speed averaging 10–15 miles per hour (16–24 kph); speeds of 35 miles per hour (56 kph) have been clocked in a few cases. The waterspout generally moves along a gently curving path. The spray ring may evolve into a spray vortex (the bush) that is raised from the surface of the water.

The decaying stage of a waterspout, which lasts up to three minutes, commences when cooler air enters the waterspout. The spiral pattern disappears, the funnel becomes distorted, and the spout disintegrates.

Vineyard Sound Waterspout: A classic waterspout formed in Vineyard Sound, an extension of Nantucket Sound, Massachusetts, in the early afternoon of August 19, 1896. It was witnessed by many residents and summer visitors on Martha's Vineyard since it lay only about 6 miles (9.7 km) offshore. Careful measurements were later made from photographs taken from several locations, and the waterspout's dimensions were calculated by triangulation. The spout height was found to be 3,600 feet (1,097 m) from water surface to cloud level, and the diameter of the funnel column was 240 feet (73 m) at the water, 144 feet (44 m) in the middle, and 840 feet (256 m) at cloud level. A large spray vortex was raised at the surface of the water to a height of 420 feet (128 m). The average forward movement amounted to only 1.1 miles per hour (1.8 kph). Three spouts formed in succession, the longest lasting 18 minutes.

Dust Devils: A dust devil is a well-developed dust whirl, not as large or vigorous as a tornado. It has many cousins, called the dancing dervish, dancing devil, devil, and satan, and, when over desert areas, the desert devil, sand auger, and sand devil. Whatever the name, a dust devil is a circular vortex of small dimensions over dry or sandy areas that carries aloft dust, leaves, and other light material picked up from the ground.

Dust devils are not only smaller and weaker than tornadoes, they also arise from quite different atmospheric conditions: strong convection during sunny, hot, calm summer days. They are generally several yards in diameter at the base, narrowing for a short distance upward before expanding again, like two cones apex to apex. Dust devils vary in height, normally rising only 100–300 feet (30–91 m); however, in hot desert country they may rise as high as 2,000 feet (610 m). Dust devils may rotate clockwise or counterclockwise, and their motion is erratic and slow as they travel from one patch of heated air to another. In desert country it is not unusual for three or more desert devils to be visible at the same time.

The mature dust devil may be divided into three vertical regions. Region 1 is a shallow, frictional layer near the ground, where air and dust spiral into the core of the whirl. Above this shallow boundary layer is Region 2, a stable vortex in which the pressure force and the centrifugal force are in balance. Region 3 begins where the top of the vortex becomes unstable and the turbulent air diffuses radially with height.

A simpler form of dust whirl, often seen at street corners, is the small, short-lived eddy caused by the meeting of winds blowing along two intersecting streets. The strong winds blowing around mountain ridges are yet another form; these whirls have been known to lift roofs and overturn parked airplanes.

Snow Devils: Vortices that form over a cover of snow are called snow devils, or snow spouts. Many of us have seen small snow devils traipsing across the snow for a few seconds before they disappear. Meteorological literature has a few references to these intriguing snow "whirlies," reported from such diverse places as Antarctica, Greenland, western Canada, Kansas, and New England. The mechanism that creates them has been the subject of great speculation among meteorologists. Although most dust devils are caused by vertical instability resulting from differential heating of a surface, these factors seem not to be present over snow surfaces. Wind eddies resulting from the presence of physical obstructions such as hills or islands have been suggested as a mechanical cause. Strong, gusty winds may be another mechanical cause, as in the cases of the relatively large snow spouts reported from Antarctica. The enormous snow spout seen near the Timpanogos Divide in Utah in 1970 was apparently a true tornado in origin, later crossing snowy terrain.

HURRICANES

A hurricane, or a tropical cyclone, is a living entity created by natural forces, with periods of gestation, birth, adolescence, maturity, decline, and ultimately death and disappearance. A hurricane's life span may be as long as 30 days or as short as a few hours. It may cross and recross wide oceans and strike at locations several thousand miles from its point of origin. The native habitats of hurricanes are the North Atlantic and eastern Pacific oceans on either side of the North American continent, but there are vigorous cousins of hurricanes, called typhoons and cyclones, in the western North Pacific Ocean and in the South Pacific and Indian oceans, respectively. Collectively, these tropical-storm disturbances have been called "the greatest storms on earth."

Definition: A hurricane is a rotating wind system that originates over warm tropical waters. It is distinguished by a calm central core or eye, within high walls of thick cloud in which the temperature is higher than in the surrounding atmosphere. Hurricanes are often accompanied by heavy-to-torrential precipitation and powerful winds, and sometimes by internal tornadic whirls. As a hurricane approaches the shallow

water of a seacoast, it causes a storm
tide to rise (to the right of the storm
relative to its forward motion),
inundating beaches and low-lying shore
areas, destroying shore installations,
and often drowning the unprepared.

Tropical Storms
of the Western
Atlantic Basin:
Most tropical storms in the North
Atlantic-Caribbean-Gulf of Mexico
realm are born out of an easterly wave
condition wherein a trough of low
pressure in the upper atmosphere
travels west, often moving from West
African waters across the entire Atlantic
Ocean to American shores. It disturbs
the resident tropical air over the warm
(at least 80°F, or 27°C) ocean waters
between 10°N and 30°N, causing
instability, i.e., rising convective
currents, cloud formation, and
thunderstorm activity.

As a result of ascending columns of air,
atmospheric pressure falls over an
expanding area. The deflective effect of
the Coriolis force (a result of the earth's
rotation) diverts the airflow into a
somewhat circular, counterclockwise
motion. Spiral cloud bands surround
the center of the storm, and in the
convective updrafts within the cloud
bands water vapor condenses and
precipitation results, releasing more
latent heat to fuel the storm system.
The rotating winds accelerate due to air
movement toward the low-pressure
center in a process known as the
conservation of angular momentum:
Wind speed must increase if the area of
rotating winds is diminished. The
innermost cloud bands form the "eye
wall," a ring of clouds extending from
the surface of the sea to great heights,
that surrounds the calm central eye. A
tropical storm with unlimited potential
and unknown destination is born.

Structure: The vertical structure of a full hurricane
reaches from the surface of the sea to
40,000 or 50,000 feet (about 7–9

miles, or 12–15 km). There is an inward flow of low-level winds below approximately 10,000 feet (2 miles, or about 3 km). In the middle section of the storm, around the central core, there is a predominantly cyclonic circulation; and above, about 25,000 to 30,000 feet (roughly 5–6 miles, or 7–9 km), high-altitude winds flow in an outward direction. The cloud walls of a hurricane may extend above 40,000 feet (about 7 miles, or 12 km), spreading out to become an anvil-shaped cloud at the base of the stratosphere and extending downwind for many miles. The whole structure, which may be up to 600 miles (about 966 km) in diameter, is steered along by the course of the winds aloft. (See the illustration of the structure of a hurricane on pages 132–133.)

If the wind speeds are 38 miles per hour (about 61 kph) or less, the storm is termed a tropical depression; if the winds range between 39 and 73 miles per hour (about 63 to 117 kph), it is called a tropical storm; a full hurricane is recognized if the winds reach 74 miles per hour (about 119 kph) or more.

Movement: Large hurricanes often develop from atmospheric disturbances, known as easterly waves, that move off the West African coast, and are carried westward across the Atlantic Ocean by the prevailing atmospheric flow. The developing storm will usually travel along the southern edge of the Azores-Bermuda High, a high-pressure zone in the mid-Atlantic that is found during the summer season between 30°N and 35°N. If the high is strong and in a normal position, the disturbance will continue westward through the West Indies and into the Caribbean Sea or the Gulf of Mexico. But if a trough of low pressure extends southward from temperate latitudes, this blocking high-

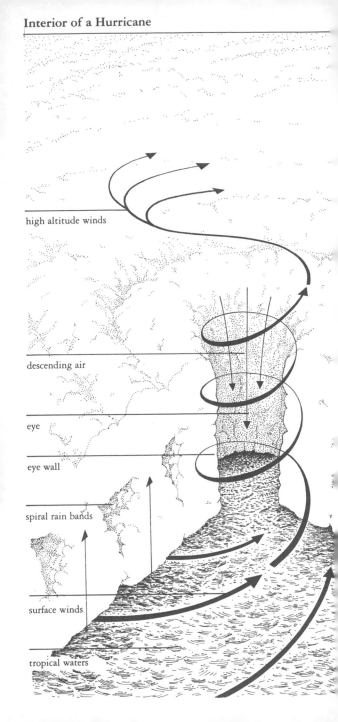

high altitude winds

descending air

eye

eye wall

spiral rain bands

surface winds

tropical waters

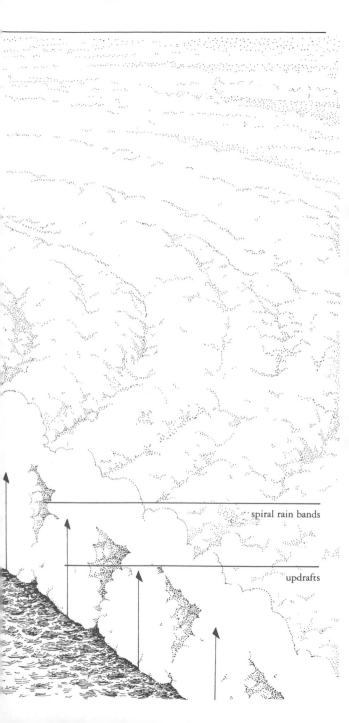

spiral rain bands

updrafts

pressure area will weaken and permit the tropical storm or hurricane to find a steering current within the trough to the northwest or north. Then the storm center will head toward the American mainland, perhaps striking the seaboard of the southeastern states or passing along the offshore waters of the North Atlantic seaboard. If the storm enters the zone of westerlies north of 30° or 35° latitude, it will move rapidly northeastward across the cool waters of the North Atlantic and lose its tropical structure.

Season: The normal tropical storm season lasts for six months, from the beginning of June to the end of November. The Gulf of Mexico and the Caribbean Sea spawn most of the early-season storms that affect the United States, and many of these storms originate from stalled late-season cold fronts that penetrate unusually far south. June storms are generally small in size and weak in intensity. Activity increases slightly in July, and storms grow in size. A marked increase in both frequency and intensity takes place in August, and the place of origin expands eastward into the broad Atlantic Ocean.
The height of the hurricane season is early September, when the tropical latitudes of the Atlantic Ocean, now warmed to 80°F (27°C), become the principal storm-generating area. One-third of all North Atlantic storms occur in September, and some of the larger developments cross the entire expanse of the Atlantic Ocean from West Africa to North America. There is a slight decrease in frequency during the second half of the month, an increase during the first two weeks of October, when the Caribbean Sea and Gulf of Mexico again become active as source regions, and then the season goes into a steady decline in late October. Few storms are charted in November.

Long-term averages show that the Atlantic normally produces ten tropical storms each year, of which six reach hurricane strength. The monthly averages are: 0.5 in June, 0.8 in July, 2.5 in August, 3.5 in September, 1.8 in October, and 0.2 in November. The 1990 season produced 14 tropical storms, of which six attained hurricane strength.

Tropical Storms of the West Coast of Central America: The warm waters of the eastern Pacific Ocean between 12°N and 25°N also provide a spawning ground for tropical storms during the warm season from June to November (the most intense of these storms usually being generated early or late in the season). The zone of activity extends from the coast of Central America and Mexico westward to the waters south of the Hawaiian Islands. On the average (based on 1977 to 1987 seasons), 21 tropical depressions form in this area annually. An average of 16 develop sufficiently to be accorded tropical storm names, and 8.3 reach hurricane intensity. The monthly frequency averages are 3.7 in July, 3.9 in August, and 2.8 in September. The 1990 season produced a total of 20 named storms, two shy of the 1987 record, and 16 were full hurricanes.

Although generally smaller in extent than their Atlantic cousins, eastern Pacific tropical storms can be equally severe over a restricted area. Some storms are capable of raising winds of 100–140 miles per hour (about 161–225 kph); extreme destruction has occurred at several Mexican ports where full-fledged hurricanes have struck. These storms are usually short-lived, as a result of their tendency to move northwest over cooler waters, where they lose their source of energy. The forward speed of movement varies from 7.5 to 18 miles per hour (12 to 29 kph), and the average duration is from four to five days, though several storms have been charted for more than ten days.

Tracks of
Tropical Storms:
Typical storm tracks move toward the west or northwest, through an ocean area rarely traversed by commercial boats (other than the tuna fleet.) Occasionally a storm may change direction, moving to the north and northeast, to strike a blow at the west coast of Mexico and Baja California. In the very active month of September 1939, this happened no fewer than five times. Five tropical storms moved northwest along the coast of Baja California; two came ashore below Ensenada, within 200 miles (about 322 km) of the United States border, and another reached 32°N before turning east to come ashore between San Diego and Los Angeles. The latter tropical storm, known as *El Cordonazo* (the Lash of St. Francis), caused approximately $2 million damage in the Los Angeles area and the loss of 45 lives, mainly at sea. In 1976, Tropical Storm Kathleen traveled north across Baja California, bringing hurricane-force gusts to Yuma, Arizona, and soaking rains as far north as Idaho and Montana. Since 1950, five storms with hurricane-force winds have passed near enough to the Hawaiian Islands to inflict damage on the Aloha State. Hurricane Iwa swept the western islands on November 23, 1982, with steady winds of 90 miles per hour (about 145 kph) and gusts of up to 110 miles per hour (about 177 kph). The center passed just west of Kauai, where total damage was estimated at $130 million; losses on Oahu were valued at $4 million.

SNOWSTORMS AND ICE STORMS

A snowstorm consists of an almost infinite number of ice crystals formed in the below-freezing environment of the middle and upper atmosphere when an abundance of moisture is present. At the core of every ice crystal is a minute nucleus, a solid particle upon which moisture condenses and freezes. In a supercooled atmosphere within a cloud, liquid water droplets and free ice crystals cannot coexist for long periods of time. The ice crystals rob the liquid droplets of their moisture and thereby grow continuously and rapidly. Some of these sizable ice crystals stick to each other to create a collection of ice crystals, known as a snowflake.

Simple snowflakes can be seen in a variety of beautiful forms. The most common is hexagonal, although the symmetrical shapes reproduced in most photomicrographs of snowflakes are not often found in actual snowfalls. Most of the snowflakes in a typical snowstorm consist of broken fragments and clusters of adhering ice crystals.

Preconditions for Snowfall: In order to sustain a snowfall, there must be a constant inflow of moisture to feed the growing ice crystals. An airstream may pick up moisture as it passes over a relatively warm water surface such as an ocean or large lake;

if the moisture is subsequently lifted to higher, cooler regions of the atmosphere, snowfall may occur. The Pacific Ocean is the source of moisture for most snowfalls west of the Rocky Mountains; the Gulf of Mexico and the tropical North Atlantic Ocean supply water vapor for potential snowfalls to the air currents over the central and eastern portions of the United States and Canada. Areas to the lee of the Great Lakes experience their own unique lake-effect storms when cold airstreams pass over warm, open-water surfaces, resulting in convection, condensation, and finally precipitation on the downwind shore. Mountainous areas, or merely rising terrain, can initiate snowfalls by orographic lifting of a moist airstream in below-freezing temperatures.

Effects of Ice Crystals: A single, minute ice crystal seems incapable of being the source of immeasurable beauty, an element of force, or a weapon of destruction, but when combined with countless others it can have a profound effect on many aspects of human life. Floating in the upper atmosphere, ice crystals cause beautiful haloes, coronas, and other fascinating optical phenomena surrounding the sun and moon. In the upper portions of cumulonimbus clouds, they contribute to the creation of electrical charges that can result in spectacular lightning and booming thunder. Falling from clouds high over the equatorial zones, snow crystals melt into large raindrops and cause the heavy downpours that nourish the rain forests. In temperate zones in winter, snowflakes fall to earth unmelted and then adhere to each other, resulting in heavy accumulations that transform landscapes into winter wonderlands, but also disrupt transportation and normal economic life. In arctic regions, they form the dreaded blizzard of the frozen polar sea and continental tundra. As they accumulate on mountain slopes, snowflakes compact and create the snowpack that is vital to the irrigation of our summer crops. But occasionally they can also be the cause of an unexpected descent of a mass of snow, called an avalanche, leaving quick death and immediate destruction in its wake. Avalanche conditions are now identified by snow hydrologists, and warnings of possible avalanches are routinely issued in several western states. Avalanches are most common when new snow settles on the crust of previous snowfalls, most often on steep mountain slopes. The descent of the snow mass also creates an "avalanche wind" before it, whose rush may uproot trees, demolish buildings, and harm people.

Snow Glossary: The following are terms used to describe variations of snow forms and snow behavior:

Snow Banner: A plume of snow blown from a mountain crest. It is sometimes mistaken for volcanic smoke, or for a cloud banner or cloud pennant.

Snow Blindness: Impaired vision or temporary blindness caused by sunlight reflected from snow surfaces. The medical term for snow blindness is niphablepsia.

Snowblink: A bright white glare on the underside of clouds produced by the reflection of light from a snow-covered surface, usually observed in arctic regions.

Snowburn: An inflammation or burn of the skin caused by the sun's rays when reflected off a snow-covered surface.

Snow Creep: An extremely slow, continuous, downhill movement of a layer of snow.

Snow Crust: A crisp, firm, upper surface on fallen snow, usually formed by melting and refreezing.

Snow Eater: Any warm wind blowing over a snow surface; usually applied to the chinook, a warm wind that flows down the eastern slopes of the Rocky Mountains.

Snow Flurry: A popular term applied to a snow shower, particularly one that is very light and brief.

Snow Garland: A rare and beautiful phenomenon in which snow is festooned from trees, fences, and other objects in the form of a rope, several feet long and several inches in diameter; it is formed and sustained by the surface tension of thin films of water bonding individual snow crystals.

Snow Grains: Precipitation in the form of very small, opaque particles of ice; the solid equivalent of drizzle.

Snow, Green: Snow surfaces with a greenish tint, a result of the presence of certain microscopic algae (cyroplankton).

Snow Line: The lower elevation of a perennial snowfield that covers an extensive mountainous area.

Snowpack: The accumulated depth of snow on the ground.

Snow Pellets: Precipitation consisting of white, opaque, approximately round ice particles having a snowlike structure. They will bounce if they fall on a hard surface. In most cases, snow pellets fall in shower form from stratocumulus or cumulonimbus clouds, often before or together with snow, and chiefly on occasions when the surface temperature is close to 32°F (0°C). They are formed as a result of accretion of supercooled droplets onto what is initially a falling

ice crystal, also called soft hail
or graupel.

Snow, Red: A snow surface that exhibits a reddish
tint as a result of the presence of certain
microscopic algae (cyroplankton) or of
particles of red dust; also called
pink snow.

Snow Roller: A mass of snow shaped somewhat like a
lady's muff, seen on rare occasion in
open fields in hilly regions. It is formed
when the snow is moist enough to be
cohesive and is set in motion by a wind
blowing down a slope. The snow mass
rolls onward until it becomes too large
(some reach four feet across and seven
feet in circumference) or until the
ground levels off too much for the wind
to propel the snow mass farther.
Before the automobile age, the term
snow roller was applied to horse-drawn
vehicles consisting of two large wooden
rollers that flattened snow on
public roads to make a smooth path
for sleighs.

Snow Shower: A brief snowfall, sometimes intense,
comparable with a rain shower.

Snowslide: Same as a snow avalanche, but usually
of smaller dimensions.

Snow Spout: A whirlwind that picks up loose snow
instead of grass and leaves; also called
snow devil. Most are formed
mechanically, by the convergence of
opposing local air currents, as
opposed to thermally (as are
whirlwinds). Some raise slabs of snow
from the surface by suction.

Snow, Spring: A coarse, granular, wet snow
resembling finely chopped ice,
generally found in spring; also called
corn snow or granular snow.

Snow, Sugar: Snow that has been recrystallized by
melting and refreezing at ground level

before spreading upward in a porous snowpack. The crystals do not adhere to one another, so the snow flows like sugar. Weak layers of this snow are the major cause of slab avalanches, in which a large crust of snow slides downhill.

Snow Tremor: A disturbance in a snowfield, sometimes called a snow quake, caused by the settling of a large area of thick snow crust over air pockets. The collapse of the snow structure may be accompanied by a loud report; over a large, level field, adjacent patches may settle in a series of tremors. Occasionally, a "snow geyser" may be blown upward through a crack in the settling crust.

Snow Warnings: Snowfalls come in all sizes and shapes, ranging from brief, intermittent flurries that never stick to the ground to continuous heavy falls that deposit a foot or more of snow. Furthermore, the flakes may lie as a uniform blanket on the earth's surface, or be blown into huge drifts separated by patches of bare ground. All these varieties of snowfall must be considered by forecasters. Following are some terms they use to distinguish among the different types of snowfall. The word *snow* in a forecast, unaccompanied by a qualifying word such as *occasional* or *intermittent,* means that the fall of snow will be steady and will probably continue for several hours.
Heavy snow warnings are usually issued to the public when a fall of 4 inches or more is expected in a 12-hour period, or a fall of 6 inches or more is expected in a 24-hour period. Some variations of these criteria may apply in different parts of the country. For example, where 4-inch snows are common, the emphasis on heavy snow is generally associated with 6 or more inches of snow. In parts of the country where heavy snow is infrequent, or in

metropolitan areas with heavy traffic, a snowfall of 2 or 3 inches will justify a heavy snow warning.

Snow flurries are light, intermittent snow showers of short duration, generally with slight accumulation. The term is a popular one, not defined in National Weather Service (NWS) literature, though sometimes NWS uses the term in forecasts. Snowfall during flurries may reduce visibility to an eighth of a mile or less.

Snowsqualls are brief, intense snow showers accompanied by strong, gusty winds, comparable to summer rainsqualls. Snowsqualls are most common in the lee of the Great Lakes and in mountainous areas.

Blowing and drifting snow generally occur together as strong winds move falling snow or loose snow on the ground. *Blowing snow* is defined as snow lifted from the surface by the wind and blown about to such a degree that a standing observer's horizontal visibility is greatly restricted. *Drifting snow* is a term used in forecasts to indicate that strong winds will blow falling or loose snow on the ground into significant piles or drifts. Drifting snow is distinguished from blowing snow in that most of the drifting snow remains within 5 feet of the ground. In the northern plains, the combination of blowing and drifting snow, without current snowfall, is often referred to as a ground blizzard.

Blizzards are the most dramatic and perilous of all winter storms, characterized by low temperatures and strong winds bearing large amounts of snow. Most of the snow accompanying a blizzard is in the form of fine, powdery particles that are whipped around in such great quantities that at times visibility is only a few yards.

Blizzard warnings are issued when winds with speeds of at least 35 miles per hour (about 56 kph) are accompanied by considerable falling and blowing snow, and temperatures of 20°F (about −7°C) or lower are expected to prevail for an extended period of time. *Severe blizzard warnings* are issued when forecasters expect wind speeds of at least 45 miles per hour (about 72 kph) along with a great density of falling or blowing snow and temperatures of 10°F (about −12°C) or lower.

Ice and Sleet Storms:
Winter's worst menace in modern times is a freezing rain, which coats outdoor objects with an icy sheath known to meteorologists as glaze. With our overwhelming dependence on electrical power for heating, cooking, communications, and even entertainment, the disruptions of power attending an ice storm can cause widespread hardship and malaise. You may escape the effects of a big snowstorm by seeking shelter inside, but the chill and discomfort of an ice storm can often be felt inside as well as outside.

Freezing rain occurs when water droplets fall from an above-freezing layer of air aloft through a shallow layer of below-freezing air at the surface of the earth. Upon impact, or shortly thereafter, the droplets freeze onto all exposed objects, coating them with an icy glaze of varying thickness. If the layer of cold air at the surface is deep, the raindrops freeze as they descend and form little ice pellets, known technically as *sleet*. These ice pellets are transparent globular or irregular grains of ice. The interior of an ice pellet may be part liquid and, in such cases, the pellet will break as it hits a hard surface. Sometimes both freezing rain and sleet occur during one storm period.

A *glaze storm* is popularly known as a silver thaw in some sections of the country, especially in the Northwest, where freezing rain often adheres to any and all outdoor objects. Despite the damage and inconvenience caused by a glaze storm, it presents a most spectacular scene when the rays of a rising sun glitter like thousands of beautiful spangles on ice-coated trees and shrubbery.

Ice generally forms on exposed objects in coatings ranging from very thin to about an inch (about 2.5 cm) thick. But deposits of up to 6 inches (about 15 cm) were reported in northwestern Texas in November 1940, and in New York State in December 1942. Sometimes a wet snow following an icing will increase the clinging accumulations to fantastic sizes. Utility poles often break down under such burdens, and many broadcast and communications towers have been bent double under their icy loads.

During Michigan's famous ice storm of February 1922, an icy covering about 12 inches (30 cm) long on a length of Number 14 telephone wire weighed 11 pounds (about 5 kg). During the memorable late-November 1921 storm that hit Worcester, Massachusetts, an evergreen tree 50 feet (about 15 m) high with an average spread of 20 feet (6 m) was supporting a burden of ice estimated to weigh 5 tons (5.5 metric tons).

HAILSTORMS

One of the surprise products of a hot summer afternoon thunderstorm is the almost magical transformation of the landscape from verdant green to icy white with the onset of a hailstorm. The first sign that hail may be arriving is a growing whitening among the shafts of rain. Soon a rattling sound is heard, as hailstones strike roofs and pavements, and the ground whitens, becoming slippery as hailstones cover grass and roadways. A hailstorm can be the most damaging part of a thunderstorm, inflicting injury on man and beast and destroying crops, gardens, and property like a giant pummeling machine.

What Causes Hailstones: A hailstone is a product of the updrafts and downdrafts that develop inside the cumulonimbus clouds of a thunderstorm, where supercooled water droplets exist. The transformation of droplets to ice requires not only a temperature below $32°F$ ($0°C$), but also a catalyst in the form of tiny particles of solid matter, or freezing nuclei. Continued deposits of supercooled water cause the ice crystals to grow into hailstones.

What we generally call hailstones have passed through several stages of accretion, from the first stage, called graupel, to small hail, to hailstones.

Sometimes only the first stage is reached; at other times hailstones from two or more stages may fall to earth simultaneously. By scientific agreement, an icy conglomeration is called a hailstone when it reaches a diameter of ⅕ inch (5 mm). In all its forms, hail usually occurs in relatively short episodes rather than as steady precipitation.

Development: The major stages in the development of hailstones can be defined as follows:

Graupel: Soft hail, or graupel, consists of white, opaque ice particles, usually nearly round (although sometimes conical), with a snowlike structure and a diameter up to ⅕ inch (5 mm). Each pellet consists of a central ice crystal that has accreted supercooled water droplets that freeze on the nuclei. Graupel is compressible and rebounds on a hard surface; thus it is sometimes called snow pellets.

Small Hail: Small hail is the same size as graupel, but differs in its higher density and partially glazed surface. Small hail particles are generally semitransparent and rounded, with conical tips and diameters up to ⅕ inch (5 mm). They consist partly of liquid water and sometimes have a frozen outer shell. Graupel transforms into small hail by the liquid water taken in through air capillaries in the ice framework.

Hailstones: Hailstones are concentrations of ice arranged in layers, with diameters greater than ⅕ inch (5 mm). Stones may be as small as peas or as large as grapefruits (about 5 inches in diameter). Usually they are roughly spherical or conical but can adapt a great variety of shapes and structures. Spherical stones, the most common form of hailstones, have a stratified interior structure somewhat resembling the rings of an onion: Layers composed of clear ice alternate with layers of white, granular ice, known as rime, formed when the hailstones are carried up and down in vertical currents within a cloud. Each layer melts a little during every descent and acquires a new sheathing of ice during each ascent into freezing temperatures, creating the onionlike layers. When hailstones are tossed out of a chimney-effect updraft into a nearby descending current, or when the supporting power of an updraft weakens, the hailstones descend to earth.

In their travels the stones acquire varying textures and appearances. Some develop protruding lobes that resemble little feet, probably as a result of spinning as they fall. Occasionally several hailstones freeze together, forming irregular chunks of ice that smash into pieces on impact with the earth.

Hailshafts and Hailstreaks: A hailshaft is a column of hail falling from a single thunderstorm cell. The ground area swept by the hailshaft is known as a hailstreak, typically produced by a hail cell (the hailshaft of a thunderstorm cell) moving along at 30–35 miles per hour (about 48–56 kph), although speeds of 60 miles per hour (about 97 kph) have been recorded. Hailstreaks normally cover areas varying from 100 feet (about 30 m) to 2 miles (about 3 km) wide and about 5 miles (8 km) long. However, they have been known to cross several counties, covering interstate tracks 200 miles (about 322 km) long. The combination of all the individual hailstreaks of a storm are known as a hailswath.

An unusually large "super-hailstreak" in Illinois in 1968 had a maximum width of 19 miles (about 31 km) and a length of 51 miles (82 km), covering 788 square miles (about 2,041 sq km). At one moment the area of falling hail measured 19 miles (about 31 km) wide by 10 miles (about 16 km) long, and the hailshaft was moving forward at 35 miles per hour (about 56 kph). Its entire duration was 90 minutes. The larger stones were 2¼ inches (about 57 mm) in diameter, and the total production was 82 million cubic feet (about 2.3 million cu m) of ice.

Hailstorm Facts: Following are some interesting facts and figures pertaining to hailstorms.

Largest Hailstone: Weather lore is full of accounts of suspiciously large hailstones—some have been reported to be the size of an elephant, while others are "merely" 20 feet (6.1 m) in diameter. In the years since scientific reporting of weather events began, however, the size of hailstones seems to have decreased considerably. For many years the largest hailstone reported and accepted by Weather Bureau officials was one that fell at Potter, Nebraska, on July 6, 1928: its circumference was 17.2 inches (43.7 cm) and its weight was 1.51 pounds (685 grams). This record was not surpassed until the "new champ" fell at Coffeyville, Kansas, on September 3, 1970. Weighing in at 1.67 pounds (758 grams), it measured 17.5 inches (44 cm) in circumference. Accounts from other areas of the world describe larger and heavier hailstones, and from the damage, injuries, and even deaths reported, the claims seem substantiated.

Record Accumulations: A severe hailstorm on June 3, 1959, at Selden, in northwestern Kansas, left an area measuring 9 by 6 miles (14.4 by 10 km) covered with hailstones to a depth of 18 inches (about 46 cm). The hail fell for 85 minutes and did $500,000 worth of damage, mainly to crops.

Hailstones tend to be swept downhill by accompanying heavy rain, eventually accumulating in deep drifts. Piles 6 feet (1.8 m) high were reported by Henry Wallace, an editor at a farm magazine, at Orient, Iowa, on August 6, 1890; some of those in protected areas remained on the ground unmelted for 26 days.

A massive hailstorm in Nodaway County, in northwestern Missouri, on September 5, 1898, left hail on the ground for 52 days, rendering ice-clogged fields unworkable for two weeks; on October 27, enough hail still remained in ravines to be used by local residents to make ice cream. On some occasions in the Great Plains, snowplows have been called out in midsummer to clear highways after a heavy hailfall.

Hail Alley: The High Plains immediately east of the Rocky Mountains experience the most frequent hailstorms in North America. "Hail Alley" extends southeast from northern Alberta, Canada, into Montana and continues southeast to include the eastern parts of Wyoming, Colorado, and New Mexico, as well as most of South Dakota, Nebraska, Kansas, Oklahoma, and west Texas. North America's most hail-prone city is Cheyenne; lying to the east of the Laramie Range in eastern Wyoming, it receives an average of 9 to 10 hailstorms per season. Some locations in the higher elevations of the Rockies may experience 20 or more hailstorms annually.

Hailstorms may occur anywhere in the United States if convective activity and sufficient moisture are present and if the freezing level aloft is relatively low. The Pacific Coast has the fewest number of hailstorms; activity increases considerably in the interior mountains. The Arctic and Tropics rarely produce hail conditions.

Death by Hail: Seemingly authentic reports of hailstorms killing people have come from around the world, notably fom China and India, whose populations are very concentrated. In northern India in 1888, hailstones as large as cricket balls (about the size of a baseball) reportedly killed 246 persons as well as 1,600 sheep and goats. Another storm, in western Hunan Province in southeastern China, was said to have killed 200 people and injured thousands in 1932.

In the United States, eight persons were reported by the *South Carolina Gazette* to have been killed by a hailstorm along the Wateree River on May 8, 1784. Other North American reports of deaths from hail came from Broome, Quebec, in 1879; Uvalde, Texas, in 1909; Windsor, North Carolina, in 1931; and near Toronto, Ontario, in 1976. Only two deaths have been authenticated by the National Weather Service (formerly the U.S. Weather Bureau). The first occurred on May 13, 1939, near Lubbock, Texas. A 39-year-old farmer died of injuries received when he was caught in an open field during a severe hailstorm. More recently, an infant lying in its mother's arms was killed by hail at Fort Collins, Colorado, on July 30, 1979.

Extraordinary Hailstorm Reports:

Eyewitness accounts of tremendous storms still make for good reading, even in this age of video recordings. Consider this report of a storm in Dubuque, Iowa, on June 16, 1882, in the *Monthly Weather Review*:

"For thirteen minutes, commencing at 2:45 P.M., the largest and most destructive hailstones fell that were ever seen at this place. The hailstones measured from one to seventeen inches in circumference, the largest weighing one pound twelve ounces [794 grams]. Washington park was literally covered with hailstones as large as lemons, and large basketfuls could be gathered in a few minutes. They exhibited diverse and peculiar formations, some being covered with knobs and icicles half an inch in length; others were surrounded by rings of different colored ice with gravel and blades of grass imbedded in them. The foreman of the Novelty Iron Works, of this city, states that in two large hailstones, melted by him, were found small living frogs. A number of persons were severely cut and bruised by the falling hailstones. The damage inflicted is estimated at $5,000 [1882 dollars]. One florist lost 2,387 panes of glass. Hundreds of windows of south and west exposure were broken, including twenty windows of heavy French glass. Railroad men report that hail fell at 2 P.M. at McGregor, forty miles to the northwest. No hail fell on the eastern side of the Mississippi [river], or at Julien, six miles west of this city."

Denver has had two very large hailstorms in recent times. The first, on June 13, 1984, lasted from 1:30 to 5:30 P.M. *Storm Data* reported:

"The worst hailstorm ever experienced in the Denver area in terms of damage battered the region for several hours. The hardest hit cities were the northwestern suburbs of Arvada, Wheat Ridge, and Lakewood, but large hail also fell in Golden, Southeast Denver, and Aurora. Damage occurred in all of these areas, but by far the worst effects of the storm were in the northwestern suburbs. Homes and other buildings sustained nearly 200 million dollars worth of damage. Many thousands of cars were battered by giant hailstones, and total damage to vehicles was estimated at 100 million dollars.

"In some areas, golfball- to baseball-size hail fell continuously for 30 to 40 minutes; some spots were pelted with a few stones as large as grapefruits. Roofs on thousands of structures were severely damaged. Uncounted car windshields were broken; two-thirds of Arvada's police cars were rendered inoperable. Torrential rain—as much as 4¾ inches (121 mm) in Lakewood—combined with the hail to clog drains and cause widespread damage from flooding. In some spots hail was washed into drifts several feet deep.

"About 20 people were injured by the giant hailstones; a couple were hospitalized. A woman drowned when she was trapped under a trailer by high water."

This 1984 storm was surpassed six years later by a storm on July 11, 1990 that caused $600 million in damage in the Denver area.

OPTICAL PHENOMENA

The sky overhead is nature's canvas. She dips into her palette and colors visions of glowing geometric designs around the sun and moon, and dancing bands of light across the heavens with all the brilliant hues of the spectrum. Many of these stunning phenomena result from the effects of solar beams passing through tiny, floating ice crystals or water droplets in the atmosphere. The striking visions of halo phenomena involve various combinations of four basic mechanisms: *reflection, refraction, diffraction,* and *scattering.*

Reflection occurs when a surface returns a portion of the incoming radiation to its original source. A mirror image is a ready example of the results of reflective action.

Refraction refers to the bending of light waves. As waves progress from a source to our eyes, their direction of motion changes (1) as the result of a difference in density within the transmitting medium, or (2) because the energy must pass through an interface between two media of differing densities.

Diffraction requires the direction of light to change as it passes by an opaque object, spreading the light around its edges. This diffraction zone is also called the shadow zone. Coronas and glories, which are discussed later in this essay, are examples of diffraction. *Scattering* is the process by which small particles, suspended in a medium, diffuse a portion of the incoming radiation in all directions. In scattering, no energy transformation takes place, only a change in the direction that light travels in space. The blue color of the sky is the result of scattering of the sun's light by small air particles.

A knowledge of the general characteristics of these four mechanisms will enable you to recognize their optical phenomena as they appear across the sky.

Halo: A halo is any one of a large class of atmospheric phenomena that appear as whitish or colored rings and arcs around the sun or moon. Haloes are seen through an ice-crystal cloud or in a sky filled with falling ice crystals. Two common types are described below.

Halo of 22°: The most commonly observed optical phenomenon is the *halo of 22°*, so-called because its radius subtends an angle of 22° from the observer. That is, it appears at an angle 22° from an imaginary line between the observer and the sun or moon. (See the illustration on page 158.) The halo is produced by refraction of light that enters one face of a six-sided ice crystal and leaves by a second face. In order for a full 22° halo to be seen, the sky must be filled with hexagonal ice crystals falling randomly, which frequently occurs. A reddish inner edge and bluish outer edge are usually the only colorations that can be discerned.

Halo of 46°: This phenomenon takes the form of a prismatically colored circle, or an incomplete arc thereof, centered on the sun or moon, with an angular radius of 46° from the viewer. The 46° angle apparently is caused by light entering one end of a crystal and exiting from one of its six sides (or vice-versa), a requirement much rarer than is needed for 22° haloes. Like the halo of 22°, it is red on the inner edge and blue on the outer and is produced by randomly oriented ice crystals.

Parhelia: Also called *mock suns* or *sun dogs,* these slightly colored luminous spots appear to the left and right of the sun at the same elevation above the horizon. Similar bright spots near the moon are known as *paraselenae.* The parhelia of 22° are produced by refraction of flat, hexagonal ice crystals falling with pricipal axes vertical. As the elevation of the sun above the horizon increases, the position of the parhelia may also increase, as far as 45° above the horizon.

Rainbow: Although sunlight appears white, it is actually a combination of all the colors of the spectrum—red, orange, yellow, green, blue, indigo, and violet—each with its particular wavelength. The rays making up this "white light" can be bent, or refracted, in various ways and can be separated into the different colors of the spectrum. For example, light rays can be refracted as they enter a droplet of water; inside the droplet they reflect off the far side of it and are further refracted as they leave. The exiting rays are returned at an angle of about 42° from the incoming rays. The result may be a rainbow.
To see a rainbow, you must view a distant area of rain while the sun's rays are passing overhead and reaching water droplets (most often the droplets are contained in a shower cloud). Some of the sky must be partially clear for the sun rays to reach the rainy area.
Each color of the rainbow we see is formed by individual light rays that reach the eye at a certain angle (the angle for each particular color never changes). The primary rainbow exhibits

Optical Phenomena

Halo of 22°

light from sun or moon

22°

halo

light from
sun or moon

22°

22°

22°

Primary and Secondary Rainbows

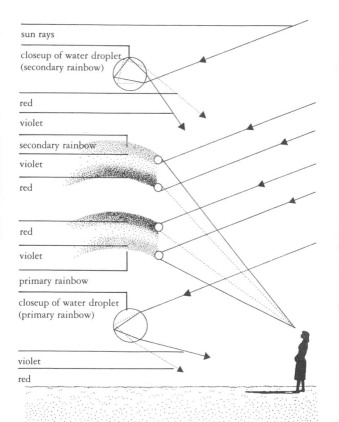

sun rays

closeup of water droplet
(secondary rainbow)

red

violet

secondary rainbow

violet

red

red

violet

primary rainbow

closeup of water droplet
(primary rainbow)

violet

red

violet and blue on the inside and yellow and red on the outside because the short waves of blue and violet are refracted more than the longer waves of red and yellow.

On some occasions, depending on the size of the rainy region and the position of the sun, you can see a secondary rainbow, less intense but with a larger radius than the primary arc and appearing above it. Secondary rainbows occur as a result of double reflections from each side of the back of the water droplets. The order of colors in the secondary rainbow is the reverse of the primary rainbow. The reds are on the inside and the blues on the outside of the colored arc.

Moonbow: Occasionally, when the moon is bright, its light can be adequately reflected and refracted to produce a colored bow, known as a *moonbow,* in the nighttime sky. It differs from a rainbow only in that a moonbow's colors are less intense. Moonbows are most frequently sighted around tropical islands (for example, Hawaii and the Caribbean), where nighttime rainshowers and partly cloudy skies are more common than at higher latitudes. Similar bows often can be observed in water spray—as grand as that of Niagara Falls, or as simple as that from the nozzle of a garden hose. Travelers may even see colored arcs in the early-morning spray across the bow of their cruise ship.

Aurora: The surface of the sun is a stormy place, with occasional disturbances that cause huge eruptions of ionized particles. Most of the material returns to the body of the sun, but a number of particles are ejected into space at speeds of several million miles per hour. Along with the particles, the flares emit short-wave ultraviolet radiation that reaches the earth in about eight minutes (the particles take about 30 hours). The magnetic fields of the earth deflect these charged particles toward the polar regions where they cause major disturbances called *magnetic storms*. When the high-speed particles strike molecules and atoms in our ionosphere, they excite atomic electrons into different orbits and cause the molecules to vibrate. When the atoms and molecules return to their original states, light is emitted. Since this process can take place over a very large region at one time, the polar skies become illuminated in spectacular displays of white or colored sheets and beams of light called auroras.

Colors: Oxygen atoms glow red or green, depending on the electron energy, and provide most of the light in a typical aurora. Nitrogen molecules also emit a red light and, at very high altitudes of several hundred miles, a violet glow. Other atoms, such as hydrogen and sodium, glow red and yellow (respectively), but their light is much fainter than the oxygen and nitrogen light. This colorful phenomenon is called the *aurora borealis* in the Northern Hemisphere and the *aurora australis* in the Southern Hemisphere, or, more popularly, the northern lights and the southern lights.

Shapes: Auroral displays take on many shapes but most often appear in the form of arcs, bands, rays, or massive curtains that seem to wave and flicker. Sometimes the lights appear to converge at the zenith in a crown, called a *corona*.

Locations: The aurora is usually produced at altitudes of about 60 to 80 miles (97 to 129 km), though it sometimes appears higher. The northern auroral zone, in which the lights appear nearly every clear night with varying intensity, extends in general from north-central Alaska through Labrador to northern Norway. On very rare occasions the aurora borealis has been seen as far south as Central America and the Caribbean Sea. On March 13, 1989, a brilliant display across southern Canada and the northern United States was seen in Guatemala and the Cayman Islands, and electromagnetically induced currents caused a power blackout in the Quebec area.

Possible connections between solar flares and meteorological occurrences on earth have been explored by many scientists. Some intriguing correlations have been suggested, but as yet no generally accepted scientific hypothesis has been developed.

FLOODS

A flood is a body of water that covers land not usually under water. Rivers are more subject to flooding than are larger bodies of water, such as lakes or arms of oceans. Some rivers overflow as a result of seasonal storms, or because of the runoff from annual snowmelt. Such added volume of water causes the water level to rise; it may eventually reach bankfull (top of the lowest riverbank). If the rise continues, flood level is reached and overflow takes place, inundating the adjacent land. Fields and crops may be covered with water, and structures damaged or destroyed by surging waters. Human life is put in jeopardy if the flood strikes at night or warnings are not heeded.

Floods in
History:
However, flooding can also be beneficial. The ancient Egyptians counted on the annual rise of the Nile River to water their crops and supply nutrients to the soil. In our western states, spring floodwaters are contained in storage reservoirs, and irrigation water is released from them during the dry season to provide a steady water supply for agricultural interests.
Four thousand years of recorded history tells of humanity's efforts to combat the devastating effects of floods. The mythologies of many different

cultures—Babylonian, Indian, Judeo-Christian, Chinese—include important flood episodes, since extensive damage can be wrought by surging high water. To get an idea of the size and force of floodwaters, keep in mind a few simple weights and measures. One gallon of water weighs about 8.5 pounds; a large bathtub holds 1 cubic yard of water, which weighs 0.75 tons. No wonder, then, that the raging waters of a flood can cause such devastation.

Human ingenuity has responded with diverse engineering attempts to tame the ravages of floodwaters, and to divert them to useful purposes.

Causes: Floods arise from many atmospheric situations—some long-term developments, others quick-breaking events. Many of the greatest inundations occur when excessive rains fall during winter storms over large river basins that are already saturated because of previous wet periods. When spring snowmelt is added to the seasonal precipitation, enormous runoffs must be handled by heavily taxed drainage systems. In the North, ice jams at critical points along a river during the spring break-up often lead to serious local flooding. In the summer and fall, tropical storms and hurricanes often carry enormous amounts of moisture north to swell coastal rivers and estuaries, adding to the height of storm-induced tidal surges. In mountain country in summertime, thunderstorms trigger cloudbursts, creating dangerous flash floods—raging walls of water that descend steep canyons at astonishing speeds and scour the hillsides of vegetation well above normal stream levels. Urban flooding results from inadequate runoff facilities or blocking of natural drainage channels by new developments and construction. Floods in the United States take an annual toll of about 100 lives and cause property damage totaling more than $1 billion (1990 dollars).

Flash Floods: Small streams confined to narrow valleys or steep canyons in mountainous country have always been subject to sudden flooding when heavy rains fall in their watershed. In the Southwest, flat areas adjacent to mountains are often hit with flash floods. When it rains heavily in the mountains, water sweeps down usually dry arroyos, and then across highways. In recent decades, such flash flooding has become the number-one weather-related killer in the United States. Real-estate development along many rivers and the increased recreational use of mountain canyons contribute to the risk of losing life and property in these flash floods.

The Greatest American Floods of the Century: North America has witnessed periods of devastating rain that have raised its rivers to the flood stage and beyond.

Great Mississippi Flood of 1927: The flood in late winter and early spring of 1927 on the Mississippi was the most devastating, in terms of lives lost and damage inflicted, of all North American river floods. Precipitation throughout the Mississippi River watershed had been more than abundant as a result of a series of storms during the autumn of 1926, and from December 1926 to April 1927 heavy rains continued through the central portions of the valley. The rains became particularly frequent and intense in March and April.

Three successive flood crests on the lower Mississippi in January, February, and April increased the magnitude of the high waters. Approximately 23,000 square miles (about 60,000 sq km) of the alluvial valley was inundated. A major storm lasting from April 12 to April 16 took place over an area of the lower Mississippi River that was already flooded, producing extremely high stages there and on the Upper Mississippi and Missouri rivers. The Ohio River was only moderately high, but the simultaneous arrival of the flow of the three rivers produced the fourth-highest stage of record at Cairo, Illinois, where the Ohio joins the Mississippi.

To the south, over the Arkansas and Red River basins and along minor tributaries, all streams swelled rapidly. Before this rise had crested, another intense storm occurred, from April 18 to April 24, over a smaller area in the center of the Arkansas and Red River basins. The Arkansas rose even more rapidly than before and resulted in the highest stage at Little Rock since 1833. The Mississippi River crested at Arkansas City on April 21. Fortunately, 23 crevasses in the main river levee caused a sharp drop after that date. If no breaks had developed, the water would have reached a considerably higher stage, probably about May 1. Without such breaks upriver, the Mississippi would have added far more to the flood flow at the mouth of the Red River.

The death toll for the 1927 flood came to 313; the damage amounted to nearly $300 million (in 1927 dollars). Some 18 million acres of land were inundated, and thousands of families had to be evacuated from their homes. At the request of the governors of the six states involved, federal assistance was provided, and Secretary of Commerce Herbert Hoover was placed in charge of the rescue work and rehabilitation.

Big Thompson Canyon Flood: On the last day of July 1976, Colorado was on the eve of its hundredth anniversary of statehood. It was a Saturday night, and the Rocky Mountain resorts were crowded with vacationers seeking relief from the midsummer heat. In the foothills east of Estes Park, where the Big Thompson

River flows through a scenic canyon, hordes of campers occupied almost every available site.

All day Saturday, cool polar air had overlain the eastern slopes of the Rockies, while out on the plains moist tropical air held sway. Although the skies were cloudy, no thunderstorm activity developed until 6:00 P.M., when a strong southeast flow carried warm, moist airstreams up over the Front Range. Thunderheads quickly formed, with tops towering to 60,000 feet (18,288 m). The storm cells became locked against the mountains over Boulder and Laramie counties, and some of the largest remained stationary over Big Thompson Canyon. Actual rainfall totals over the mountains will never be known, but radar estimates ranged as high as 10 inches (254 mm) falling within 90 minutes.

Severe flooding began shortly after 7:00 P.M. The waters crested and destroyed the small settlement of Drake, at the river's junction with its North Fork, at about 9:00 P.M. The water velocity was extreme—as high as 25 feet per second (7.6 m per second). A wall of water estimated at 19 feet (about 6 m) high swept tremendous amounts of debris downstream, including large trees and several structures.

The flash flood took at least 139 lives in its sweep of the Big Thompson canyon; several more people were reported missing. Most campers had no warning of the wall of water about to descend on them. A survey reported 323 houses and 96 mobile homes destroyed. The Loveland municipal plant, a brick building, was lifted from its foundations; much additional damage resulted in towns on the plains beyond the canyon's mouth. Total losses were estimated at $30 million.

DROUGHT

Few things are as devastating to the human race as a drought. Many North Americans live in areas of marginal water supply, where they sometimes receive adequate rainfall, but at other times experience a serious deficiency. On the Great Plains, once known as the Great American Desert, ingenuity has partially subdued this tendency to drought so that today, in most seasons, the area yields a bounteous harvest of crops. Still, the threat of drought and crop failure is always present.

Droughts occur when atmospheric currents fail to carry moisture to an area. For instance, when high pressure overlies the central valley of the United States, airstreams from the Gulf of Mexico are prevented from reaching the area. Southwest winds from the southwestern desert and Mexico prevail and inhibit precipitation. In summertime, the Azores-Bermuda high-pressure area moves westward and prevents Atlantic Ocean and Gulf of Mexico moisture from reaching the southeastern states, causing drought conditions.

An extended period of moisture deficiency may cause a serious hydrologic imbalance, resulting in crop damage, reduced water supplies, and economic disruption. Dry spells occur sporadically in all parts of the United States and Canada, but the term *drought* is reserved for dry periods that are relatively extensive in both time and space—generally, a period of 21 days with only 30 percent of normal precipitation. A moisture shortage that is termed a drought in one region may not be considered so elsewhere, and a shortage may be less serious in one season than it would be in another.

Drought Severity Index: A sophisticated index of meteorological drought was developed in the 1950s by Wayne Palmer of the National Weather Service. The Palmer Drought Severity Index is based on the concept of supply and demand: Supply is represented by precipitation and stored soil moisture; demand is figured by a formula that combines moisture loss through evapotranspiration (evaporation from land and water surfaces and transpiration from vegetation) with both the amount of moisture needed to recharge the soil moisture and the amount of runoff required to keep rivers, lakes, and reservoirs at normal levels.

The results of this water-balance accounting produce either a positive or a negative figure, which is weighted by a climatic factor. The final figure is an index that expresses the degree of abnormality over a period of several months in a particular place. An index of + 2 to − 2 indicates normal conditions; − 2 to − 3 indicates a moderate drought, − 3 to − 4 a severe drought, and below − 4 an extreme drought.

Crop Moisture Index: A second, more short-term, index measures the degree to which growing crops have received adequate moisture during the preceding week. The Crop Moisture Index is computed from average weekly values of temperature and precipitation. Taking into account the previous moisture conditions and current rainfall, the index determines the actual moisture loss and hence the crop's current moisture demand.

If moisture demand exceeds available supplies, the Crop Moisture Index has a negative value; if current moisture meets or exceeds demand, the value is positive. Charts of the Drought Severity Index and the Crop Moisture Index for the United States are published weekly during the growing season in the *Weather and Crop Bulletin,* issued jointly by the U.S. Department of Agriculture and the National Weather Service in Washington, DC.

Dust Storms: During periods of drought, unusual and frequently severe disturbances called *dust storms* can arise, most frequently on the Great Plains. (Dust storms are also popularly called *dusters* and *black blizzards.*) Strong winds sweep up very small grains or particles of earth and raise them as a cloud high into the air. Winds generally must exceed 20–25 miles per hour (32–40 kph) to accomplish this. A dust storm sometimes arrives suddenly in the form of an advanced haboob or dust wall, resembling a front, which may be many miles long and several thousand feet high. It may be preceded by one or more dust whirls or dust devils (see the essay on tornadoes), either detached from or merging with the main mass. The advancing dust may penetrate window frames or small cracks in houses and can cause respiratory distress in human beings and animals.

In American weather terminology, dust that reduces visibility to between $\frac{5}{16}$ and $\frac{5}{8}$ mile (0.5–1 km) is reported as a dust storm; if the visibility is reduced below $\frac{5}{16}$ mile (0.5 km), the condition is reported as a severe dust storm. Dust may be carried to great heights and remain suspended in the air for many days. A single dust storm may cover several states or move out over water; dust from the Great Plains, for example, has been spotted over the Atlantic Ocean.

Historic Droughts:
Drought has been a significant influence on North American history. Drought threatened the crops planted by the Pilgrims in Massachusetts during their first summer in 1621, when no rain fell for 24 days. Another dry spell, in 1623, endangered their corn crop. During the severe droughts of 1749, 1761, and 1762 in eastern New England, fields of grain and many homes caught fire. Very dry summers occurred in 1805 and 1822 in the Eastern states, and the summer of 1854 brought widespread drought and consequent crop failures to areas from New York to Missouri. In newly settled Kansas, the year 1860 was notorious for drought, searing heat, and dust storms. After some 20 years of varying but mostly adequate rainfall on the Great Plains, drought struck in 1881 and continued until 1887, when a combination of severe winters and near-drought summers—compounded by overpopulation of animals—caused the collapse of the Cattle Kingdom. Subsequent growing seasons were inconsistent in yield, and economic distress on the Great Plains was widespread, prompting an exodus of people from Kansas and Nebraska. A historian of the period has written: "Fully half the people in western Kansas left the country between 1888 and 1892. Twenty well-built towns in that part of the state were reputed to have been left without a single inhabitant." An oft-repeated phrase of the times was, "In God we trusted, in Kansas we busted." Conditions became even worse when extreme drought, a low of -4.97 on the Drought Severity Index, prevailed in 1893, 1894, and early 1895.

Droughts of the Twentieth Century: Droughts threatened to reach serious levels again in 1910, 1911, and 1913, but as it turned out they were relatively short-lived.

The greatest disaster in American history attributable to meteorological factors occurred on the Great Plains in the 1930s, a period popularly known as the Dust Bowl Days. Drought conditions developed in the latter half of 1930 and continued into the first half of 1931, when most of the northern and southern Great Plains experienced a severe moisture shortage. Every year thereafter until the end of the decade, some part of the vast region was affected by serious drought; in 1934 and 1936, the entire region from Texas to Canada suffered. Between August 1932 and October 1940, the Drought Severity Index in western Kansas was continuously below normal, and, after 38 months, extreme drought prevailed. Low points reached the extreme category three separate times: −5.96 in August 1934, −5.55 in August 1936, and −5.55 again in October 1939.

The term "Dust Bowl" applied only to the most seriously affected areas of the south-central Great Plains, which included parts of Colorado, New Mexico, Texas, Oklahoma, and Kansas. At its greatest extent, in the winter of 1935–36, the Dust Bowl covered 50 million acres. For each of the preceding three years, Dalhart, Texas, close to the center of the area, had an average of 11.08 inches (281 mm) of rainfall, or 58 percent of normal. The stricken area began to shrink in 1937, and by early 1938 it was down to about 9.5 million acres (about one-fifth its original size), before more extreme drought in 1939 expanded it somewhat. Significant rains finally came in 1940 and early 1941, putting the area back into agricultural production just in time to meet the extraordinary demands World War II

placed upon the United States'
productive capacity.

The following account was written
by Richard R. Heim, Jr. for
Weatherwise magazine:

"The 1940s brought generally wet
weather to most of the nation, but
widespread drought returned in the
1950s. During 1953 and 1954, severe
drought stretched from the Southeast
across the Ohio Valley and into the
central Rockies and Great Basin, at its
peak covering 51 percent of the
country. The dry area expanded in
1956, covering the southern and central
Great Plains. In sections of Texas the
first seven months of 1957 were the
driest since 1934. At Amarillo in
northwestern Texas the worst drought
of record prevailed from mid-1952 to
early 1957, when annual precipitation
was only 61 percent of normal, and
dust storms once again swirled across
the Great Plains.

Generally speaking, the United States
has been in an overall wet spell since
the 1950s. Indeed, eight of the twenty
wettest years for the country as a whole
occurred in the 1970s and 1980s,
according to records going back to
1895. (In comparison, the 1930s had
six of the twenty driest years.)

In spite of the overall wetness, regional
droughts did occur in the 1960s and
1970s. The northern Rockies were dry
during 1960 and 1961, as was much of
the Midwest in 1963 and 1964.
Moderate drought appeared in the
Northeast in 1961, then worsened into
the extreme category during 1964 to
1966. In New York City, reservoirs
dropped to about 30 percent
of capacity.

Severe drought returned briefly in the
Southwest and the Southern Plains in
1971 and again in 1972 and 1974.
Extreme drought occurred during 1967
and 1977 in the Far West, especially in
California, and also in the Northern

Plains Great Lakes region. For some of the people in these areas, the 1976 and 1977 drought was the most severe of this century.

Drought appeared briefly during 1980 and 1981 in scattered areas across the country. The drought of 1988, however, was the major one of the 1980s. During the summer, the Mississippi River reached near-record low levels at many gauging stations, restricting barge and towboat traffic. At Memphis, the river level was the lowest since 1872, the year record-keeping for Memphis began. Reservoir levels in California were also down, though not as low as in the 1970s. Farmers in the Plains and the Midwest were suffering what *U.S. News and World Report* called 'potentially the worst crop disaster since the 1930s.' Parts of the Southeast, too, were experiencing the most devastating drought of the century."

Recent Droughts: By mid-July of 1988, about 45 percent of the country was suffering severe-to-extreme drought. The combination of searing heat and meager rainfall in 1988 cut U.S. production of corn by one-third, of soybeans by one-fifth, and of spring wheat by more than one-half. Drought again restricted the wheat production on the Great Plains. Kansas suffered a lack of rainfall in early 1989 that diminished the winter wheat crop. North Dakota suffered all summer from extraordinary heat and received only about 50 percent of normal rainfall. The spring wheat crop suffered irreparable damage.

During 1990, drought continued to make headlines in three major agricultural areas. In California, paltry winter precipitation again depleted water supplies. Winter precipitation was below normal for the fourth consecutive year—less than one-half of normal precipitation in several parts of the state. Early in 1991, restrictions on urban and farm use of water were put into operation. In the Dakotas, January 1990 was the warmest January since 1881, and the year as a whole was among the driest of the century. Spring and summer rains relieved the topsoil dryness, but did little to lessen the effects of long-term drought prevailing since the fall of 1987. Florida and much of the Southeast endured a long-term drought beginning in September 1988. By the end of February 1990, the drought was considered the worst in southeastern Florida since 1895. Rain came in spring and summer of 1990 to bring major relief to the region.

Part II
Color Plates

Key to the Color Plates

The color plates on the following pages
are divided into 16 groups:

The Atmosphere
Cirrus Clouds
Middle Clouds
Stratus Clouds
Cumulus Clouds
Orographic Clouds
Mixed Skies
Showers and Thunderstorms
Tornadoes and Other Whirls
Hurricanes and Tropical Storms
Snowstorms and Ice Storms
Floods
Drought and Related Events
Obstructions to Vision
Optical Phenomena
Historical Photographs

Thumb Tab Guide: Each of these groups is represented by a symbol, a graphic silhouette that conveys the kind of meteorological phenomena shown within the group. On the pages that follow, you will find a table showing all of the symbols used to represent each of these groups. On the right is a list of weather events and their plate numbers—the photographs included within each particular group.

Symbol	Subjects	Plate Numbers
The Atmosphere	spacecraft and satellite views	1–11
	radar images	12, 13
Cirrus Clouds	cirrus	14–22
	cirrocumulus	23–26
	cirrostratus	27–32
	contrails	33, 34
Middle Clouds	altocumulus	35, 37–48
	altostratus	36, 49–54
Stratus Clouds	stratus	55–61
	stratocumulus	62–67
	nimbostratus	68–75

Symbol	Subjects	Plate Numbers
Cumulus Clouds 	small cumulus	76, 80–88
	swelling cumulus	77–79, 89–99
	cumulonimbus	100–127, 129, 130
	gust fronts	128, 131, 196, 197, 199
	mammatus	132–135, 194, 195
Orographic Clouds 	cumuliform	138–141
	stratiform	136, 137, 141–149
Mixed Skies 		150–166
Showers and Thunderstorms 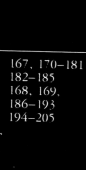	thunderstorms	167, 170–181
	hailstorms	182–185
	lightning	168, 169, 186–193
	mammatus, gust fronts, squall lines, and microbursts	194–205

Symbol	Subjects	Plate Numbers
Tornadoes and Other Whirls 	tornadoes and funnel clouds mesocyclones waterspouts, dust devils, and steam devils	207, 209–214, 223–230 206, 215–222 208, 231–236
Hurricanes and Tropical Storms	hurricanes and tropical storms viewed from space	237–241 242–255
Snowstorms and Ice Storms 	snowfalls, snowstorms, avalanches, and blizzards ice storms frost and frost patterns	256–266, 283, 284 267–274 275–282
Floods 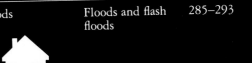	Floods and flash floods	285–293

The color plates on the following pages
are numbered to correspond with the
numbers preceding the text accounts.

THE ATMOSPHERE

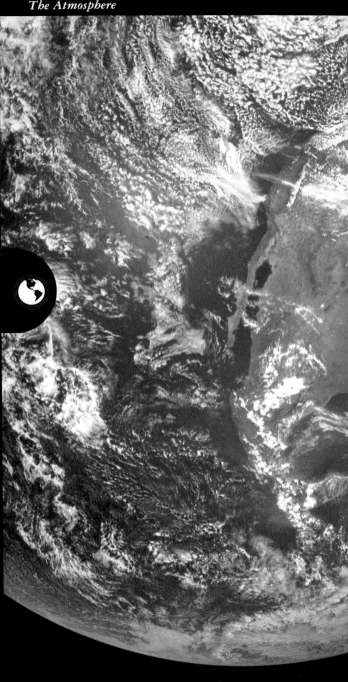

1   Earth from Apollo 16, *p. 435*

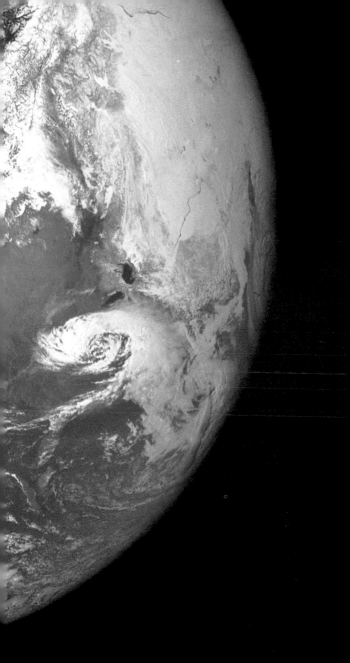

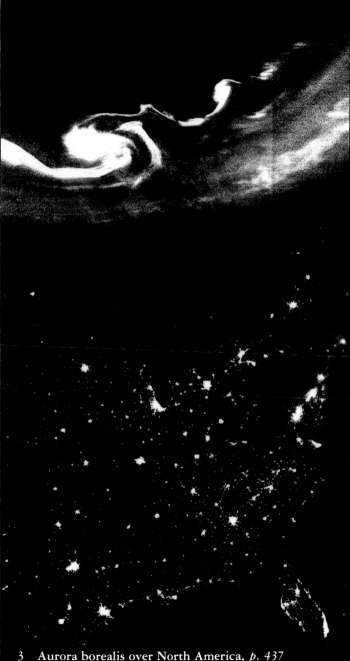

3   Aurora borealis over North America, *p. 437*

4   Frontal system: satellite view, *p. 438*

5   The Northeast: satellite view, *p. 439*

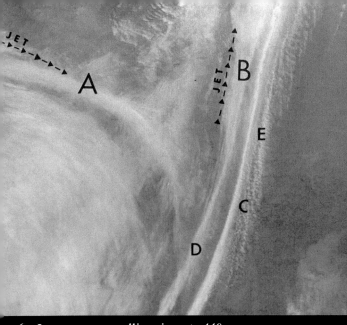

6 Jet streams: satellite view, *p. 440*

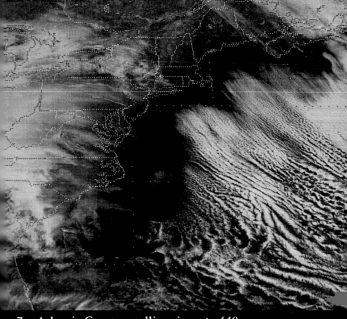

7 Atlantic Coast: satellite view, *p. 440*

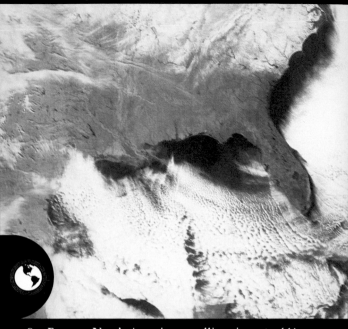

8   Eastern North America: satellite view, *p. 441*

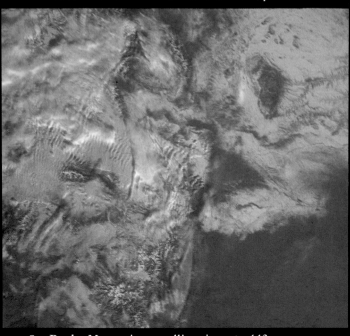

9   Rocky Mountains: satellite view, *p. 442*

10   Fog in California's Central Valley, *p. 443*

11   Eastern Pacific: satellite view, *p. 443*

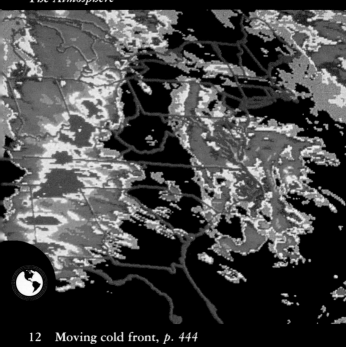

12   Moving cold front, *p. 444*

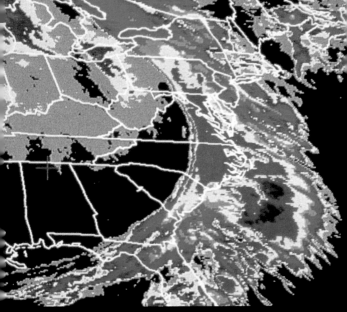

13   Moving cold front, *p. 444*

CIRRUS CLOUDS

14 Cirrus castellanus, *p. 446*

15 Cirrus intortus, *p. 446*

16 Cirrus fibratus, *p. 446*

17 Cirrus: Kelvin-Helmholtz waves, *p. 446*

18 Cirrus, *p. 446*

19   Cirrus uncinus: mares' tails, *p. 446*

20   Cirrus uncinus: mares' tails, *p. 446*

21   Cirrus radiatus, *p. 446*

22   Cirrus fibratus, *p. 446*

23   Cirrocumulus undulatus, *p. 447*

24   Cirrocumulus undulatus, *p. 447*

25   Cirrocumulus undulatus, iridescence, *p. 447*

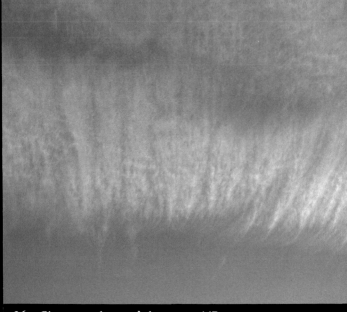

26   Cirrocumulus undulatus, *p. 447*

27   Cirrostratus nebulosus, *p. 449*

29   Cirrostratus nebulosus, *p. 449*

30   Cirrostratus fibratus, *p. 449*

31 Cirrostratus duplicatus nebulosus, *p. 449*

32 Cirrostratus nebulosus with halo, *p. 449*

33 Contrails, *p. 450*

34 Contrails, *p. 450*

MIDDLE CLOUDS

35   Altocumulus duplicatus, *p. 452*

36   Altostratus, *p. 454*

37   Altocumulus undulatus, *p. 452*

38   Altocumulus undulatus, *p. 452*

39   Altocumulus undulatus, *p. 452*

40   Altocumulus undulatus, *p. 452*

41 Altocumulus: mackerel sky, *p. 452*

42 Altocumulus castellanus, *p. 452*

43 Altocumulus: mackerel sky, *p. 452*

44 Altocumulus: mackerel sky, *p. 452*

45   Altocumulus, *p. 452*

46   Altocumulus duplicatus, *p. 452*

47 Altocumulus: mackerel sky, *p. 452*

48 Altocumulus duplicatus *p. 452*

49   Altostratus, *p. 454*

50   Altostratus, *p. 454*

51   Altostratus undulatus, *p. 454*

52   Altostratus, *p. 454*

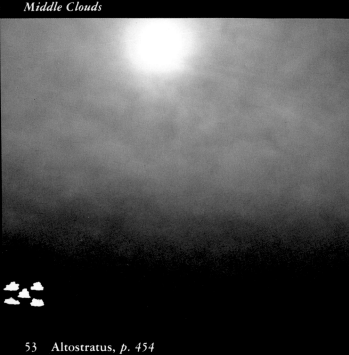

53 Altostratus, *p. 454*

54 Altostratus, *p. 454*

STRATUS CLOUDS

55   Stratus, *p.* 457

56   **Stratus undulatus,** *p. 457*

57   **Stratus opacus nebulosus,** *p. 457*

58   Stratus opacus uniformis, *p.* 457

59   Stratus translucidus, *p.* 457

60   Stratus, *p.* 457

Stratus, *p.* 457

62   Stratocumulus undulatus opacus, *p. 459*

63   Stratocumulus undulatus, *p. 459*

64   Stratocumulus, *p. 459*

65   **Stratocumulus,** *p. 459*

66   Stratocumulus undulatus, *p. 459*

67   Stratocumulus undulatus, *p. 459*

68 Nimbostratus opacus, *p. 462*

9 Nimbostratus praecipitatio, *p. 462*

70   Nimbostratus opacus, *p. 462*

71   Nimbostratus praecipitatio, *p. 462*

72   Nimbostratus, *p. 462*

73   Nimbostratus praecipitatio, *p. 462*

74   Nimbostratus with scud (pannus), *p. 462*

75   Nimbostratus, *p. 462*

CUMULUS CLOUDS

76 **Cumulus humilis,** *p. 464*

77 Cumulus congestus, *p. 466*

'8   Cumulus congestus with rain, *p. 466*

79   Cumulus congestus with rain, *p. 466*

80 Cumulus humilis, *p. 464*

82 Cumulus humilis fractus, *p. 464*

83 Cumulus mediocris, *p. 464*

84   Overcast sky changing to cumulus, *p. 464*

85   Cumulus series: 2 of 4, *p. 464*

86 Cumulus series: 3 of 4, *p. 464*

87 Cumulus series: 4 of 4, *p. 464*

88 Cumulus mediocris, *p. 464*

89 Cumulus congestus, *p. 466*

90   Cumulus congestus, *p. 466*

91   Cumulus congestus, *p. 466*

92 Cumulus congestus, *p. 466*

Cumulus congestus with rain, *p. 466*

94 Cumulus congestus, *p. 466*

95 Cumulus congestus, *p. 466*

96   Cumulus congestus, *p. 466*

97   Cumulus congestus, *p. 466*

98 Cumulus congestus with rain, *p. 466*

99 Cumulus congestus: lake-effect clouds, *p. 466*

100 Cumulonimbus calvus, *p. 469*

101 Cumulonimbus calvus, *p. 469*

102   Cumulonimbus with virga, *p. 469*

103   Cumulonimbus with virga, *p. 469*

104   Changing cumulonimbus with pileus, *p. 469*

105   Cumulonimbus series: 2 of 4, *p. 469*

108   Changing cumulonimbus, *p. 469*

109   Cumulonimbus series: 2 of 3, *p. 469*

110 Cumulonimbus series: 3 of 3, *p. 469*

111 Cumulonimbus *p. 469*

112 Cumulonimbus, *p. 469*

114   Cumulonimbus calvus, *p. 469*

115   Cumulonimbus, *p. 469*

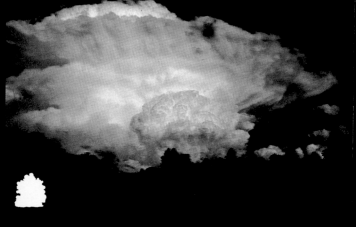

116 Cumulonimbus, *p. 469*

Cumulonimbus, *p. 469*

118   Cumulonimbus, *p. 469*

119   Cumulonimbus incus, *p. 469*

120   Cumulonimbus, *p. 469*

umulonimbus, *p. 469*

122   Cumulonimbus incus, *p. 469*

Cumulonimbus, *p. 469*

124   Cumulonimbus incus, *p. 469*

25   Growing cumulonimbus, *p. 469*

126 Cumulonimbus series: 2 of 3, *p.* 469

127 Cumulonimbus series: 3 of 3, *p.* 469

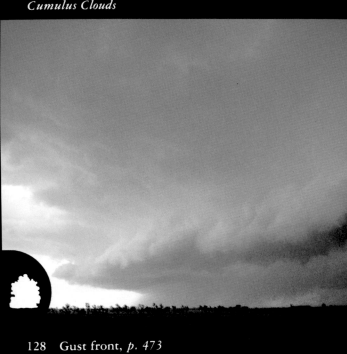

128 Gust front, *p. 473*

129 Cumulonimbus, *p. 469*

130 Cumulonimbus, *p. 469*

131 Gust front, *p. 473*

132   Mammatus, *p. 472*

133   Mammatus, *p. 472*

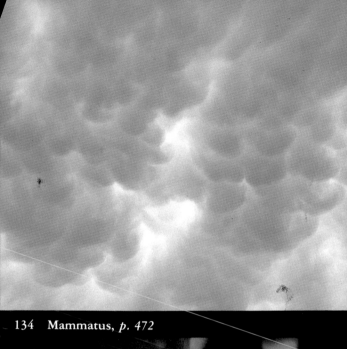

134 Mammatus, *p. 472*

OROGRAPHIC CLOUDS

136   Sierra wave, *p.* 477

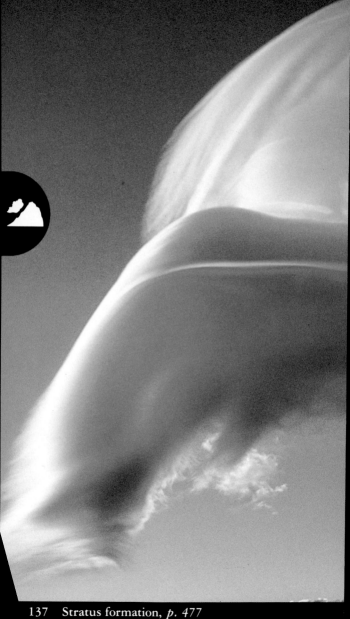

137 Stratus formation, *p.* 477

138   Cumulus formation, *p. 476*

139   Cumulus formation, *p. 476*

140 Cumulus formation, *p. 476*

141 Cumulus and stratus formation, *p. 476*

142   Stratus formation: lenticularis, *p. 477*

143   Stratus formation, *p. 477*

144 Stratus formation, *p.* 477

145 Stratus formation, *p.* 477

146   Stratus formation capped by pileus, *p.* 477

147   Kelvin-Helmholtz waves, *p.* 477

148   Stratus formations, *p.* 477

149   Stratus formations, *p.* 477

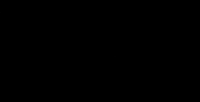

150 Altocumulus and cirrus, *p. 480*

151 Cirrus and cumulus, *p. 481*

152 Cumulus congestus, altocumulus *p. 481*

153   Cumulonimbus, cumulus, altocumulus, *p. 482*

54   Cumulus, altocumulus, cirrus, *p. 483*

155 Cumulus, altocumulus, cirrus, *p. 484*

156 Altocumulus, altostratus, cirrostratus, *p. 4*

157 Orographic cumulus and cirrus, *p. 486*

Cumulus and cirrus, *p. 487*

159 Cirrus forms, *p.* 487

160 Cumulonimbus: strato- and altocumulus

161 Altocumulus, cirrus, and orographic, *p. 490*

Cumulus and cirrus, *p. 491*

163 Cumulus and cirrus, *p. 492*

164 Cumulus, altocumulus, and cirrus, *p. 492*

165   Cumulus and stratus, *p. 494*

166   Cumulus, cumulonimbus, cirrus, *p. 495*

SHOWERS AND
THUNDERSTORMS

`7 Thunderstorm, *p. 497*

168   Cloud-to-ground lightning, *p. 502*

170   Thunderstorm, *p. 497*

172 Thunderstorm, *p. 497*

173 Thunderstorm, *p. 497*

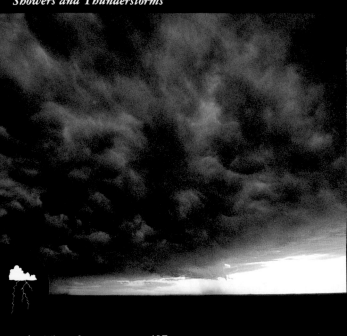

174   Thunderstorm, *p. 497*

derstorm, *p. 497*

176 Thunderstorm, *p. 497*

178 Thunderstorm, *p. 497*

Thunderstorm, *p. 497*

180 Thunderstorm, *p. 497*

181 Thunderstorm, *p. 497*

182   Thunderstorm with hail, *p. 497*

Thunderstorm with hail, *p. 497*

184 Thunderstorm with hail, *p. 497*

185 Thunderstorm with hail, *p. 497*

186   In-cloud lightning, p. 502

ive flash lightning, p. 502

188 Cloud-to-air lightning, *p. 502*

189 Positive flash lightning, *p. 502*

Showers and Thunderstorms

190 Cloud-to-ground lightning, *p. 502*

Cloud-to-ground lightning, *p. 502*

192   In-cloud lightning, *p. 502*

193   Cloud-to-ground lightning, *p. 502*

194 Mammatus, *p. 472*

5 Mammatus, *p. 472*

196   Gust front, *p. 473*

197   Gust front, *p. 473*

Showers and Thunderstorms

198   Squall lines, *p. 506*

pp. 473, 506

200   Squall lines, *p. 506*

201   Squall lines, *p. 506*

202   Developing microburst, *p. 508*

Microburst series: 2 of 4, *p. 508*

204   Microburst series: 3 of 4, *p. 508*

205   Microburst series: 4 of 4, *p. 508*

TORNADOES AND OTHER
WHIRLS

208   Dust devil, *p. 520*

209   Developing tornado, *p. 513*

215   Mesocyclone, *p. 511*

216   Mesocyclone, *p. 511*

217 Mesocyclone, *p. 511*

218 Mesocyclone, *p. 511*

219   Mesocyclone, *p. 511*

220   Mesocyclone with incipient tornado, *p. 511*

221 Mesocyclone, *p. 511*

222 Mesocyclone, *p. 511*

223 Funnel cloud, *p. 513*

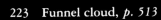

224 Funnel cloud, *p. 513*

225   Tornado, *p. 513*

226   Funnel cloud, *p. 513*

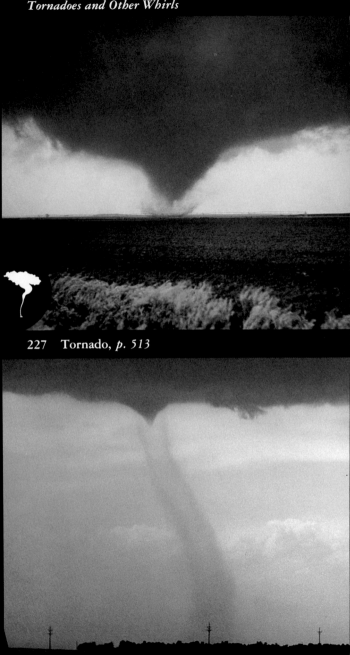

227   Tornado, *p. 513*

229 Non-supercell tornado, *p. 513*

230 Non-supercell tornado, *p. 513*

231   Waterspout, *p. 518*

232   Waterspout, *p. 518*

233 Dust devil, *p. 520*

234 Dust devil, *p. 520*

235   Steam devil, *p. 520*

236   Steam devil, *p. 520*

rricane Allen, *p. 523*

238 Tropical storm, *p. 523*

240   Hurricane Frederic, *p. 523*

241   Hurricane Gilbert, *p. 523*

242   Typhoon Pat, *p. 527*

243   Hurricane Elena, *p. 527*

244 Typhoon Gladys, *p. 527*

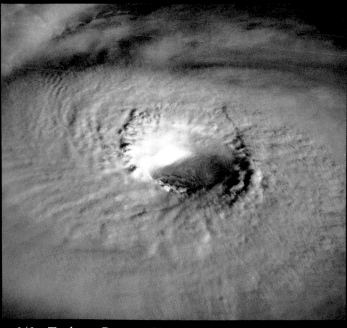

245 Typhoon Pat, *p. 527*

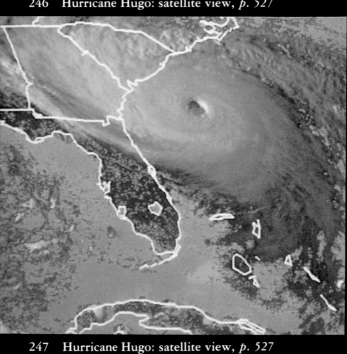

246   Hurricane Hugo: satellite view, *p. 527*

247   Hurricane Hugo: satellite view, *p. 527*

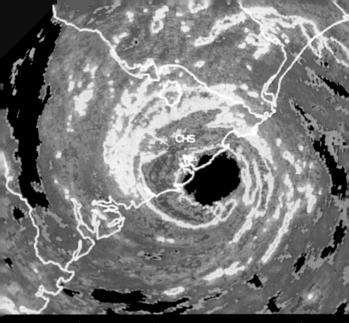

248   Hurricane Hugo: radar image, *p. 527*

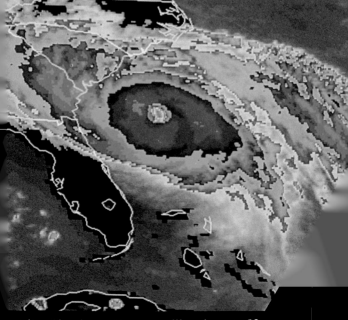

249   Hurricane Hugo: satellite view, *p. 52*

250 Hurricane Roslyn develops, *p. 527*

252   Hurricane Roslyn series: 3 of 4, *p.* 527

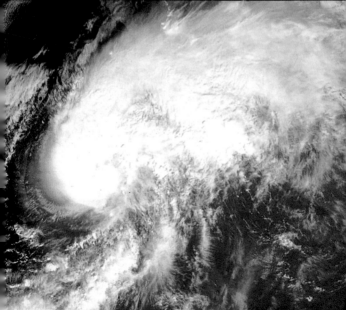

253   Hurricane Roslyn series: 4 of 4, *p.* 527

254 Hurricane David, *p. 527*

255 Hurricane Diana, *p. 527*

SNOWSTORMS AND ICE STORMS

vfall, *p. 530*

259 Desert snowfall, *p. 530*

ime and fog from thermal spring, *p. 530*

261   Snowfall near a thermal creek, *p. 530*

262   Snowfall with subsun, *p. 530*

263 Snowfall in mountains, *p. 530*

264 Snowstorm over water, *p. 530*

265 Drifting snow, *p. 530*

266 Ground blizzard, *p. 530*

267 Frozen spray from lake, *pp. 530, 534*

, *pp. 530, 534*

269   Pack ice, *pp. 530, 534*

270   Pancake ice, *pp. 530, 534*

271   Glaze ice, *pp. 530, 534*

n spray, *pp. 530, 534*

273 Glaze ice, *pp. 530, 534*

274 Icicles, *pp. 530, 534*

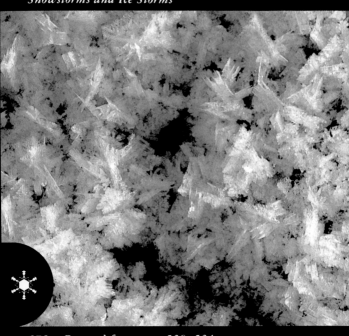

275   Ground frost, *pp. 530, 534*

st, *pp. 530, 534*

277   Air bubble in ice, *pp.* 530, 534

278   Newly formed ice pattern, *pp.* 530, 534

279   Hail, *pp. 530, 534*

round frost, *pp. 530, 534*

281   Ground frost, *pp. 530, 534*

282   Rime, *pp. 530, 534*

283   Windblown snow, *p. 530*

Wind-sculpted snow, *p. 530*

FLOODS

286   Flood, *p. 538*

288   Flood, *p. 538*

289   Flood, *p. 538*

290   Flash flood, *p. 538*

Flooding river, *p. 538*

292   Flash flood, *p. 538*

294 Everglades: dry season *p. 541*

295   Drought: dessicated vineyard, *p. 541*

Drought: effect on reservoir level, *p. 541*

297   Dust storm, *pp. 541, 544*

298   Cracked mud, *p. 541*

299 Cracked mud, *p. 541*

301 Sand dunes, *p. 541*

302 Dust storm, *pp. 541, 544*

303 Forest fire, *p. 546*

305   Forest fire, *p. 546*

306   Forest fire, *p. 546*

307 Forest fire: regrowth after damage, *p. 546*

308 Brush fire, *p. 546*

OBSTRUCTIONS TO VISION

309   Radiation fog, *p. 563*

310   Valley fog, *p. 563*

311   Advection fog, *p. 563*

312 Valley fog, *p. 563*

313 Coastal advection fog, *p. 563*

314 Coastal advection fog, *p. 563*

315 Advection fog, *p. 563*

316 Radiation fog, *p. 563*

317 Coastal advection fog, *p. 563*

318   Pollution: unnatural color, *p. 566*

319   Pollution and its sources, *p. 566*

320   Pollution and its sources, *p. 566*

321   Brown cloud of pollution, *p. 566*

322 Pollution and its sources, *p. 566*

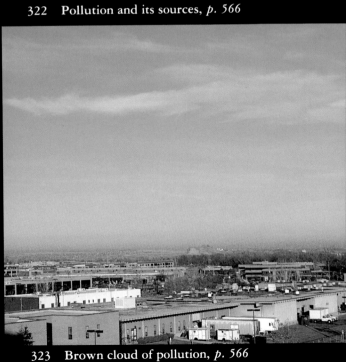

323 Brown cloud of pollution, *p. 566*

324　Clear visibility, *p. 566*

325　Visibility obscured by pollution, *p. 566*

326 Dust and haze, *p. 566*

328   Aurora borealis: curtains, *p. 548*

329   Rainbow with supernumerary, *p. 558*

331   Corona, *p. 555*

332   Corona, *p. 555*

333   Iridescence, *p. 556*

334   Iridescence, *p. 556*

335   22° halo, *p. 551*

336   22° halo, tangent arcs, *p. 551, 553*

337 Parhelia, parhelic circle, *p. 552*

338 9°, 18°, 20°, 24°, and 35° haloes, *p. 551*

339   Halo, parhelic circle, and arcs, *pp. 551–554*

340   Circumzenithal arc, *p. 553*

341   Glory, *p. 556*

342   Subsun, *p. 554*

343 Crepuscular rays, *p. 557*

344 Crepuscular rays, *p. 557*

345   Sun pillar, *p. 554*

346   Sun pillar, *p. 554*

347   Rainbow with secondary bow, *p. 558*

348   Rainbow, *p. 558*

349   Rainbow, *p. 558*

350   Rainbow with secondary bow, *p. 558*

351　Rainbow, *p. 558*

352　Rainbow, *p. 558*

353 Rainbow, *p. 558*

354 Fogbow, *p. 558*

355 Volcanic twilight, *pp.* 557, 560

356 Twilight, *p.* 560

357   Green flash, *p. 561*

358   Green flash, *p. 561*

359 Aurora borealis: emerging arc, *p. 548*

360 Aurora borealis: curtains, *p. 548*

361   Aurora borealis: corona, *p. 548*

362   Aurora borealis: rays, *p. 548*

363   Aurora borealis: rays, *p. 548*

364   Aurora borealis: curtains, *p. 548*

HISTORICAL PHOTOGRAPHS

365 Before Hurricane Camille, *p. 571*

366 After Hurricane Camille, *p. 571*

367 Windstorm damage, *p. 571*

368 Long-term wind damage, *p. 571*

369 Dust storm, *p. 571*

` of drought, *p. 572*

371   Surging floodwaters, *p.* 572

372   Flood damage, *p.* 572

373　Heavy snowfall, *p. 572*

'4　Glaze-storm damage, *p. 573*

375　Snowfall accumulation, *p. 573*

376　Snow rollers, *p. 573*

377   Severe floodwaters, *p. 573*

378   Severe floodwaters, *p. 574*

Part III
Text Accounts

The numbers preceding the text
accounts on the following pages
correspond to the plate numbers in the
color section.

THE ATMOSPHERE

Meteorologists use satellite images of
the earth to study and predict weather
patterns. Photographs taken from far
overhead make it possible to trace the
course of storms, jet streams, cloud
formations, and other developments
that influence our weather because they
show weather patterns over vast
distances, thus revealing conditions
in one area that will soon affect the
weather in other parts of the continent.

1 Earth from *Apollo 16*

View: The earth as seen from *Apollo 16,* taken
from over the Tropic of Cancer two
hours after launch.

Date: April 16, 1972.

Description: The photograph shows almost all of the
United States and much of Canada. The
large white expanse at top right is the
polar ice cap and snow-covered tundra;
the Gulf of California is easily
identified at the center of the image.
Part of northern Mexico and the
southern United States are clear of
clouds, as are the western Great Lakes,
where the ice cover is visible. The spiral
cloud pattern of a cyclonic storm covers
the northeastern United States, while
western Canada and the mountains of
the western United States are covered
with snow. A southern jet appears off
California, and a northern jet, visible at
the center of the image, covers part of
the North American land mass.
Cyclonic activity can be seen in the
northwest, southwest of Alaska. Clouds
in the middle lower section are most
likely cumulus variations induced by
high mountain chains in Central and
South America.

Significance: Photographs taken from space enable
one to see and interpret a wide range
of meteorological phenomena. This

example shows areas of cloudiness, clear weather, snow cover, ice, cyclonic activity, and locations of the jet streams—data from which unfolding weather activity can be predicted. The satellite photo was taken about two hours after the launch of *Apollo 16,* one of the series of manned flights to the moon.

2 West Coast: Satellite View

View: Satellite view of the western coast of North America from 510 miles (820 km) above the earth.

Date: July 15, 1984.

Description: Easily distinguished at the right center of the image is Vancouver Island. At the top is Alaska and the Aleutian chain; the southernmost point in the image is Point Arena, just north of the San Francisco Bay area. To the west, over the Pacific Ocean, is a large low-pressure system. The white lines embedded in the clouds, resembling contrails, are made by ship traffic on the ocean.

Significance: Under certain favorable conditions, such as fairly high humidity, water vapor condenses around the hot particulate matter pouring forth from a ship's smokestack. The condensation forms distinct and persistent trails that are often visible even from space.

3 Eastern North America: Satellite View

View: Satellite view of eastern North America by night.

Date: October 11, 1985.

Description: At the top of the photograph are the colorful swirls of the northern lights, or aurora borealis. Across the middle of the image one can see the lights of Montreal, Toronto, Buffalo, Cleveland, Detroit, and Chicago, with Minneapolis-Saint Paul to the northwest. The Boston-New York-Philadelphia-Baltimore-Washington corridor appears as a line of illumination from northeast to southwest. In the south, Atlanta, Dallas and Fort Worth, and Houston stand out, and both coasts of the Florida peninsula are outlined by city lights.

Significance: A clear night provides a superb opportunity for viewing (and photographing from space) the aurora borealis—northern lights—over Canada, and the bright electric urban lights of the United States. The flashing, colorful display of the aurora borealis (aurora australis in the Southern Hemisphere) is caused by charged particles emitted by the sun, called the solar wind. Upon entering the atmosphere, these are diverted to the magnetic poles of the earth, causing the energized atoms of oxygen and molecules of nitrogen to glow (like a neon light). The occurrence of auroras heightens with increased sunspot activity, which induces magnetic storms that disrupt radio and telegraphic communications. (The effects of sunspots on overall weather patterns are subjects of continued study.)

The aurora occurs in the atmosphere in myriad forms at altitudes of 60–100 miles (97–161 km) above the earth.

4 Frontal System: Satellite View

View: A panorama of a frontal system and attendant cloud structure, as seen by a meteorological satellite.

Description: In this image the main front runs from location *A* to location *D,* separating polar air over Canada from tropical air over the United States. *A,* marked by pointed indicators, denotes a cold front that is advancing from the northwest to the southeast, with clouds both preceding and following the frontal line, where precipitation (of the shower type) usually occurs. Convective clouds with scattered showers often follow the front.

At position *B,* rounded indicators mark the location of a warm front on the surface of the earth. Warm air from the south is overrunning the cold polar air at the surface, causing steady precipitation. Increasing moisture is being fed into the system the farther west one progresses.

At point *C,* the warm front *B* joins a new cold front, *D,* that is emanating from a new cyclonic system. The stretch from *D* to *C* is an occluded front, in which the new cold front has overtaken the warm front. Turbulent clouds have formed in and near this frontal zone. The cold front extends south of the juncture and is preceded by considerable cloud cover, but clearing takes place closely behind it.

E is the warm sector of the cyclonic system, where tropical air is moving northward in the circulation. Some convective clouds have formed in the western part of this warm sector.

Significance: Such a scene is typical of the organization of a cyclonic system over the eastern half of North America. Along the frontal boundary, clouds cover the sky and precipitation often takes place either in advance of or in the vicinity of the frontal positions.

5 The Northeast: Satellite View

View: Meteorological satellite photograph of northeastern North America from an altitude of 22,000 miles (35,400 km).

Date: Winter season.

Description: This plate shows eastern Canada and the northeastern United States, including the entire Great Lakes region. The western shores of Lakes Superior, Michigan, and Huron are visible, while the central and eastern parts of the lakes are covered with clouds. Likewise, the surfaces of Lakes Erie and Ontario are obscured by cloud cover. Bands of heavy clouds (A), containing snow showers, stream from the eastern ends of the lakes.

Significance: A predominantly westerly circulation illustrates the lake-effect snow-shower activity that is a unique feature of the climate of the Great Lakes. In late autumn and early winter, when cold air from Canada blows across the open waters of the still-warm lakes, convection takes place. Moist air rises until its dew point is reached; then condensation occurs and stratocumulus clouds form. These clouds carry heavy loads of moisture that is not released until the airstream reaches the hills lining the lake shores, when it is forced to ascend.

Comments: In winter, with its below-freezing temperatures, much of the precipitation around the Great Lakes takes the form of concentrated snow showers. These often occur in long, narrow bands reaching several miles inland. The snowflakes are of light density but often accumulate to remarkable depths of 12–24″ (30–60 cm) in short periods. The snow continues to fall as long as the wind blows with sufficient strength and from the proper direction for snow making. By midwinter, when the lakes usually freeze over, the lake-effect mechanism ceases to function.

6 Jet Streams: Satellite View

View:	Jet streams as seen from the satellite *Polar Orbiter.*
Description:	The image shows jet streams at different altitudes. The stream of cirrus *A* is a normal, high-level westerly jet flowing from the northwest until it turns to the south near the center of the picture. *B* is another high-level jet stream. Both are accompanied by deformation and shear zones. *D,* a jet at a lower level flowing from the southwest, is probably a component of the subtropical jet. The cirrocumulus clouds between *C* and *E* are the generating convective cells of cirrus streamers, which are distorted into transverse lines and streaks as they fall.
Significance:	Jet streams—swift-flowing, narrow ribbons of air that meander through the middle latitudes of both hemispheres—are vital determinants in the circulation of the upper atmosphere. The main westerly jet in the Northern Hemisphere pursues a generally easterly course between 40°N and 50°N and separates polar air to the north from tropical air to the south. Along the meeting place of the airstreams, barometric depressions develop and move generally eastward, causing cyclonic storms. In wintertime, a subtropical jet stream flows from the central North Pacific Ocean west of Mexico across the southern United States, introducing vast quantities of tropical moisture to fuel our winter storms.

7 Atlantic Coast: Satellite View

View:	Satellite image of the Atlantic seaboard under a high-pressure ridge.
Date:	February 1981.
Description:	The image focuses on a ridge of high pressure extending from the province of

Quebec to the West Indies between 65°W and 80°W. In the center of a high-pressure area, air descends and dissipates cloudiness; thus, here, because it is close to the high-pressure ridge, almost all the coastline from western Nova Scotia to southern Florida is clear of clouds and visible. To the west and east of the ridge, air is ascending in troughs of low pressure and causing cloudiness. South and southeast of Cape Cod is a retreating high-pressure area with a strong northwesterly airflow, marked by convective clouds arranged in streets from northwest to southeast. To the west, the Appalachians and Great Lakes lie under the control of an advancing trough of low pressure. Westerly and southwesterly airflows are bringing stratus clouds over the rear of the departing high-pressure ridge.

Significance: High-pressure ridges, or anticyclones, and low-pressure troughs, or cyclones, pass in an endless parade across North America, causing alternating zones of fair and stormy weather.

8 Eastern North America: Satellite View

View: Satellite photograph of eastern North America.

Date: March 3, 1980.

Description: The eastern United States is half covered with clouds and half in the clear. The northeastern Gulf of Mexico is cloud free, although most of the remainder is under clouds. Along the southern fringe, the details of the landscape are highlighted by the presence of snow.

Significance: This photograph, taken immediately after the great arctic outbreak into the South on March 1–2, 1980, illustrates the differing effects of a cold air mass on land and on water. At that time temperature at Miami dropped to freezing—the lowest recorded reading known so late in the season—and a heavy snowfall occurred from northern Alabama to New Jersey, measuring at its heaviest, in eastern North Carolina, to as much as 25″ (63.5 cm). Cold, unstable air flowing off the land has caused cloud streets to form in the eastern Gulf of Mexico west of Florida, while the cold air over the southeastern states has stabilized, yielding almost cloudless conditions there. The western Gulf of Mexico and western Caribbean Sea are covered by a solid cloud of another wind circulation of different airstreams. Near the top of the photograph, unfrozen parts of the Great Lakes are visible.

9 Rocky Mountains: Satellite View

View: Satellite view of the Rocky Mountains from 510 miles (820 km) above the earth.

Date: January 16, 1986.

Description: In among the snow-capped peaks of the Rocky Mountains, lenticular cloud formations can be discerned, indicating the presence of lee waves.

Significance: When air passes over a boundary or barrier such as a mountain ridge (in this case, the Rockies), disturbances in the airflow result. These disturbances, known as "lee waves" or "mountain waves," are stationary with respect to the barrier. Lee waves also exist in streams and rivers, and can be detected as ripples in water flowing over a large rock. In the atmosphere, the presence of lee waves over a mountain is often confirmed by almond-shaped lenticular clouds, which are generated orographically and remain nearly stationary with respect to the terrain. Lee waves may occur up to 10 miles (16 km) apart.

10 Fog in the Central Valley

View: California's Central Valley from 510 miles (820 km) above the earth.

Date: December 24, 1978.

Description: This photograph shows the huge Central Valley of California—comprising the Sacramento and San Joaquin valleys—blanketed in an extensive, thick, low-level fog. Snow can be seen to the east on the peaks of the Sierra Nevada range and, to the west, cirrostratus and cirrocumulus clouds are visible.

Significance: When warm, humid air passes over cold ocean waters, it is cooled and its relative humidity rises. By late afternoon and through the night, radiation cooling takes place, augmenting the cooling effects of the ocean and further lowering the temperature of the air mass. As a consequence, the atmospheric humidity of the warm air approaches saturation, and condensation occurs. The huge blanket of fog in this image is known as valley fog. Cool, moist, maritime air has been carried into the Central Valley by onshore winds from the Pacific Ocean. During the night the cool air radiates heat upward and cools further. Where the cool air and warm air meet aloft, fog forms and often persists through the daytime. Fog of this type may continue for as long as a week, keeping the valley floor shrouded in fog while the upper hills on each side of the valley bask in sunshine.

11 Eastern Pacific: Satellite View

View: Satellite image of the Pacific Ocean west of Baja California.

Date: September 8, 1979.

Description: Guadalupe Island, Mexico, lies near the center of the image at about 29°N; it is located about 180 miles (290 km) off the coast of Baja California and 300 miles (480 km) south-southwest of San Diego. The Gulf of California is prominent in the upper right. The Bay of Sebastian Vizcaino, which marks the dividing line between Baja California Norte and Baja California Sur, can be seen, partially hidden by clouds, at right center.

Significance: The photograph shows eddy currents in the low-level wind field. The wind direction, from the southeast, creates a cloudless half ring around Guadalupe Island and a large open space downwind. Clouds off the mainland of Baja California are fog or stratus, and well offshore, a ripple effect is visible in the cloud sheets over the ocean, perhaps indicating weak convection from warmer ocean water. A line of cumulus clouds overlies the Sierra San Pedro Martir mountains, which form the backbone of Baja California Norte.

12, 13 Moving Cold Front

View: Enhanced satellite view of a cold front above the eastern United States on successive days.

Dates: (12) 2:30 P.M., CST, November 15, 1989; (13) 9:00 A.M. , CST, November 16, 1989.

Description: The photographs show a well-developed cold front that extends from the Great Lakes to the Gulf of Mexico as it moves across the eastern United States.

Significance: Here it is possible to see the strong thunderstorms that developed along this front, especially in its southern portion, during an afternoon and evening in mid-November 1989, when as many as 39 tornadoes were reported. The most powerful of these touched down southwest of Huntsville, Alabama, at 4:30 P.M. on November

15, plowing through a business and residential area in the southern part of the city and causing 21 fatalities and 463 injuries. Property damage there alone amounted to $100 million. Alabama endured 7 tornadoes and Georgia 5 while the front passed. The very high winds that followed the cold front resulted in the collapse, on November 16, of a school wall in New York State, killing 9 children.

CIRRUS CLOUDS

High-level clouds (those generally occurring at 16,500'/5,000 m and higher over most of the United States) made of ice crystals that appear as wispy, fibrous strands stretching across the sky are called cirrus clouds. Patches of thin, white cirrocumulus clouds, easily identified by their consistently arranged ripples, signal moisture at high levels.

The virtually featureless cirrostratus, often accompanied by other types of cirrus clouds, also indicate a significant amount of moisture at high altitudes. Contrails are straight lines of ice-crystal clouds formed by water droplets emitted by aircraft. The water droplets freeze into crystals and leave behind narrow condensation trails in the atmosphere's upper level.

14–22 Thin Cirrus
Cirrus castellanus, fibratus, radiatus, spissatus, uncinus, intortus

Description: *Thin, white to light gray ice-crystal clouds,* often in the form of streamers, not covering the entire sky. Cirrus clouds are often elongated in several separate segments in the same direction across the sky.

Environment: Thin cirrus clouds form when there is moisture at high levels colder than freezing; they occur very frequently in one form or another. Cirrus clouds of some kind usually accompany organized storm systems, but such clouds are often hidden by lower clouds.

Season: All year.

Range: Wispy cirrus clouds are common all across North America. In the drier southwestern United States, they are quite frequently observed, mainly because the passing weather systems do not form intervening lower clouds that would block them from view.

Variations: Fibratus (16, 22), the most common cirrus clouds, have a striated composition. Uncinus (19, 20) are hooked forms, often called mares' tails, that emanate from a denser portion of the cloud (18, 22). Radiatus (21) have long, parallel streamers of relatively similar size that appear to converge toward the horizon. Castellanus (14) rise from a thicker base of cirrus into a turreted shape, and indicate instability at the cirrus level. Intortus (15) have very irregular shapes. Very rarely, cirrus cloud tops show a full wave breaking (17). Such a Kelvin-Helmholtz (K-H) wave lasts only a few seconds to a few minutes and indicates a large change in wind speed over small increases in altitude.

Significance: Isolated cirrus clouds are common and are usually of no great significance. Castellanus may indicate instability in the area, especially if they form toward midday in warm months in a region without much other cloud cover. An increasingly dense and widespread coverage of fibratus or radiatus is often seen ahead of advancing weather systems, and can indicate a change toward a more active weather situation. This indication applies best in the colder half of the year in the westerly-wind areas of the northern United States and Canada.

Comments: Cirrus clouds that are more or less isolated are quite common whenever there is some moisture in the upper levels. In most cases, they indicate a relatively neutral situation.

23–26 Cirrocumulus
Cirrocumulus undulatus

Description: Generally *thin white clouds*, most often composed of ice crystals, with *small but regular ripples*, usually not covering the entire sky. Cirrocumulus clouds often occur simultaneously with other cloud types, generally forming at high levels.

Environment: Patches of cirrocumulus represent moisture at high levels colder than freezing. They generally occur in areas with above-average moisture or instability at the cirrus level. Cirrocumulus clouds are sometimes hidden by lower clouds, such as those that accompany storm systems, but they may often be seen from an aircraft.

Season: All year; never very frequent or widespread.

Range: Cirrocumulus clouds are often mixed with other layered clouds at middle and upper levels and can be seen all across North America.

Variations: The characteristic ripples of undulatus are almost a necessity for identification of a cloud as cirrocumulus (24). The most important visible clue for identifying cirrocumulus is the small size of these waves (23–26). This feature distinguishes it from altocumulus, which has a similar shape but has much larger individual elements and waves. A transition into a smoother cirrus form is evident in 23; a secondary ripple pattern is present in 26. Iridescence (25) indicates that the cloud consists of water droplets of uniform size.

Significance: A small or isolated cirrocumulus cloud is not especially important, but several patches or a widespread layer may indicate instability in the area, and may be moving ahead of an advancing weather system. Cirrocumulus clouds are found with larger-scale weather systems more often than they accompany local showers and thunderstorms.

Comments: Cirrocumulus clouds are relatively infrequent, and their presence should prompt a more careful watch of the sky for other signs of increasing moisture or instability.

27–32 Cirrostratus
Cirrus duplicatus, fibratus, nebulosus

Description: *Uniform*, generally *featureless, thin to thick, white or light gray ice-crystal clouds* at the same high altitude as cirrocumulus, covering part or all of the sky. No streamers or other identifiable features are seen within most of the cloud. Haloes may be visible, especially when the cirrostratus is so thin it is not otherwise visible.

Environment: Cirrostratus clouds represent a significant amount of moisture at higher levels colder than freezing. When cirrostratus cover much or all of the sky, there are often accompanying thin cirrus or cirrocumulus, as well as clouds at other levels, as moisture flows overhead from a larger-scale disturbance. Often hidden by lower clouds, some cirrostratus clouds usually accompany organized storm systems. In the western United States, cirrostratus may stream overhead from the south or west, because of condensation upstream that is occurring only at upper levels.

Season: Mostly winter, but possible any time.

Range: Cirrostratus clouds are most widespread in the northern states, because they accompany the large-scale moving weather systems common there. Cirrostratus are visible in the southwestern United States in the monsoon of late summer, and in the southeastern United States, where they flow outward from the upper levels of subtropical and tropical disturbances.

Variations: The most common feature of cirrostratus is a vague nebulosus appearance (27, 29, 31, 32). Sometimes cirrostratus clouds are clearly a single layer of uniform cloud thickness (29). They may also tend toward fibratus (28, 30) or show their ice composition by a halo (32). Sometimes cirrostratus clouds appear to have more than one layer, in which case they are called duplicatus (31).

Significance: A layer of widespread cirrostratus
clouds may be indicative of an
approaching active disturbance. It also
may represent an inactive high-level
moisture outflow, such as during the
monsoon in the southwestern United
States; these clouds also form over the
southeastern United States when
moisture flows outward from the tops
of tropical or subtropical disturbances.
If cirrostratus clouds are becoming
more dense and are increasing with
time, other clouds should also be
watched, because their behavior
may indicate a more active
weather situation.

Comments: Although not spectacular in
appearance, cirrostratus clouds are
important because they indicate
large-scale moisture that is advancing
or has moved in from a moister region
upstream. Watch for an increase in
other clouds over a period of several
hours, indicating a disturbance on
the way.

33, 34 Contrails

Description: A *straight, narrow line of ice-crystal clouds*
produced by flying aircraft.
Condensation trails (contrails) may
occur in several short segments at
various angles to each other, depending
on flight paths.

Environment: Contrails are produced in nearly
saturated conditions at higher levels
colder than freezing. Sometimes, the
contrails are formed by several
planes flying at different altitudes,
indicating moisture through a deep
layer of the upper atmosphere. Flying
aircraft emit water droplets into the
atmosphere. The droplets immediately
freeze into crystals (the temperature is
typically colder than $-25°F/-32°C$);
these crystals last much longer in the
sky than water droplets. Since jet

aircraft can fly at a wide range of altitudes, it is not possible to identify the cloud height from visual clues alone. Contrails occur quite commonly, but they are often hidden by lower clouds.

Season: All year.

Range: All locations.

Variations: A newly formed contrail (33) is generally thinner and smaller than an older one (34). If the air is very dry, the contrail may be exceptionally thin. When the air at the aircraft's level is near saturation, a contrail may last a long time (34). When they start to become diffuse, contrails can spread to form a major cloud cover (34).

Significance: The presence of a few isolated contrails is quite common and is of no great significance. The amount of air traffic over a region also strongly influences the occurrence of contrails. An increased number of contrails, together with portions of other stratiform clouds moving into an area, may indicate a local atmospheric change due to an advancing weather system.

Comments: If the atmosphere is very moist, it will already be filled with clouds at the height of the contrail, so any contrails present may be obscured.

MIDDLE CLOUDS

Middle-level clouds form a major portion of active weather systems moving across the country. The air at these altitudes (approximately 6,500–16,500'/2,000–5,000 m) is generally stable and without vertical currents. The two principal types of middle-level clouds are altocumulus and altostratus. Altocumulus are mostly composed of water droplets and appear as textured bands or patterned grayish-white sheets. An increase in thickness and area of altocumulus clouds signals a rise in moisture at cloud level. Altostratus are composed mainly of water droplets and are organized in gray to blue sheets that have little texture. An increase of altostratus clouds may be a harbinger of more widespread precipitation in the region, though it may not happen directly overhead.

35, 37–48 **Altocumulus**
Altocumulus castellanus, duplicatus, stratiformis, undulatus
Stratiform Clouds

Description: *Thin or thick, white to gray, mainly water-droplet clouds* at middle levels with significant variations across the sky. The appearance of altocumulus varies within the cloud itself, and may change almost completely across the sky.

Environment: Altocumulus clouds form when there is moisture present at middle levels, and when temperatures range from somewhat colder than freezing to a little warmer than freezing. Larger areas of altocumulus usually accompany the more important moving weather systems and the flow of significant moisture over hundreds of miles. Other layered clouds, such as altostratus, cirrostratus, cirrocumulus, and stratus, may occur at the same time. Altocumulus clouds may also exist with cumuliform clouds of all types.

Season: All year.

Range: Altocumulus may occur anywhere, and for any of a wide variety of reasons. During the winter in the northern states and provinces, altocumulus clouds occur near or not too far ahead of (within 100–200 miles/about 160–320 km) moving weather systems. In the summer in the northern states, they may also be produced by these systems as well as by nearby large thunderstorm complexes. Altocumulus is also produced across the southern states by northward outflow from tropical and subtropical systems south of the United States, or from moisture flowing northward over the western and Plains States from the southwestern monsoon. In the West, where low-level cloudiness is much less frequent, owing to lower humidity near the ground, altocumulus is seen more often than in humid regions.

Variations: Altocumulus clouds are often seen as rows of undulatus (37-40). At other times, there is a patterned and organized look to the clouds, the mackerel sky of sailors' lore (41, 43, 44, 47). The sun can be seen through the cloud (41, 47) or be hidden (46). Occasionally, a castellanus form (42) occurs indicating instability in the region, usually in the morning during the warm season. Often, a multilayered, more disorganized, and variable duplicatus sky (35, 46, 48) is seen. Very infrequently, the mainly water-droplet form of altocumulus converts to ice particles when the cloud is much colder than freezing (45).

Significance: Altocumulus that last only a few hours or cover only a small portion of the sky sometimes form without an organized weather system or flow of moisture into the area. When altocumulus cover a larger area for a longer period of time, a more significant source of moisture is present. If cloud layers at other levels also exist, there is the possibility of a more organized weather system at, near, or approaching that area. If the altocumulus are growing in area and thickness with time, and are approaching from the wind direction at the cloud's level, then a change toward moister weather is indicated. The total sky view will help in assessing the significance of the altostratus overhead.

Comments: If altocumulus are below 6,500' (about 2 km), they are technically considered to be stratocumulus.

36, 49–54   Altostratus
Altostratus undulatus, uniformis
Stratiform Clouds

Description: *Thin or thick, gray to pale blue, mainly water-droplet clouds at middle levels* without significant variation across the sky or within the cloud itself. The precise height of completely featureless stratiform clouds is sometimes difficult to determine—the important feature is that fewer differences are visible within altostratus than in lower stratus clouds, because of the former's greater distance from the ground. For this reason, a stratiform cloud that appears to be very smooth is more likely to be altostratus than stratus. (If there is much variation, the cloud is an altocumulus.) An altostratus cloud lacks the wisps or long streamers of cirrostratus, but an indistinct veil of precipitation may be falling from its base, making it look diffuse.

Environment: Altostratus indicate moisture at middle levels with temperatures that range from somewhat colder than, to a little warmer than, freezing. An organized weather system or wind flow is needed to produce altostratus clouds, because the cloud layer must undergo prolonged gradual lifting until saturation occurs. The cloud material often is carried over horizontal distances of hundreds of miles without significant changes.

Range: Altostratus may occur anywhere in the United States and Canada. In the winter in the northern states and Canadian provinces, altostratus clouds occur far ahead of or near (100–200 miles/about 160–320 km) traveling synoptic disturbances. In the summer, they may also be produced by such systems in the northern states and Canada. Altostratus may also be produced by outflow at upper levels from large nighttime convective weather systems that occur over the center of the United States during warmer months. Altostratus may be produced across the southern states by northward outflow from tropical and subtropical systems to the south of the United States, or from moisture flowing northward over the western states and Plains States from the southwestern monsoon.

Variations: The sun may be dimly visible through a thin, uniform, and very diffuse altostratus layer (49, 52, 53), or the clouds may be totally without features and very thick (54). Note the small cumulus below the altostratus in plate 52; otherwise, the cloud cover is unvarying across the sky. Somewhat regular variations (undulatus) may occur (51), or the cloud matter may be spread out in long rows (36) and have no particular pattern at all. The clouds may also have a more varied appearance (50, 53). The cloud in plate 50 is altostratus because it is above some lower stratocumulus visible on the horizon, yet cannot be high enough to be cirrostratus because it lacks streamered features.

Significance: Any significant portion of the sky covered with altostratus clouds is cause to try to identify the source of the moisture. If the flow is from the southwest during the summer in the Four Corners states (Utah, Colorado, Arizona, New Mexico), for example, a significant amount of moisture may be over the area, and there is a possibility of heavy rain or other thunderstorm activity. In any region, if the altostratus coverage is increasing with time, and is approaching from the direction of the wind at the altostratus' level (for example, increasing from the west with west winds aloft in the winter in the northern states), then a change toward moister weather is likely. Altostratus may look more extensive toward the horizon than overhead (36, 52), but the apparent compression of cloud cover toward the horizon may be more a function of perspective than of a real increase in clouds.

Comments: Altostratus clouds are hard to identify and not particularly spectacular in many cases; however, if the sky is watched carefully over a period of time, they will usually show an important feature to distinguish them from other cloud types. If these clouds are increasing with time, a major change toward more widespread precipitation may be expected in the region, although it may not happen directly overhead. In some cases, it is not very important to distinguish a layer of altostratus thickening with time from stratus or thick cirrostratus, since all three indicate a local trend toward increasing moisture.

STRATUS CLOUDS

Layered water-droplet clouds that form in a gray layer close to the ground (below 6,500'/2,000 m) are called stratus (from the Latin word for "layer"). Stratus clouds usually yield no precipitation other than drizzle, ice crystals, steady rain, or snow grains. Their structure is so undefined that they resemble fog, except stratus clouds are above the ground.

Stratus clouds occur in humid air as a result of the cooling of the earth's surface; the flow of moist, cold air into a region at low altitudes; or the transition from fog to cumulus. Stratocumulus clouds are formed when stratiform clouds—whose shape develops horizontally—coexist with heaped clouds. Stratocumulus appear as large white and gray patches. Nimbostratus are thicker and darker than other types of stratus clouds and bring with them various types of precipitation.

55–61 **Stratus**
Stratus opacus, translucidus, undulatus, uniformis
Stratiform Clouds

Description: *Low-altitude, light to dark gray water-droplet clouds without significant variations in appearance* across the sky. Stratus clouds show little variability in appearance within the clouds, and generally no precipitation falls from them except a few drops of drizzle or small snowflakes. Since stratus clouds are quite close to the ground, they appear to have large but vaguely outlined structures; if the cloud appears very smooth, it is more likely to be an altostratus. The base of a stratus may be as low as only 100' (30 m) above the treetops or as high as 6,500' (1,981 m);

above that height such clouds are technically considered altostratus.

Environment: Stratus indicate low-level moisture at temperatures ranging from somewhat above freezing to well below it. Stratus can form locally overnight as a result of the cooling of the earth's surface; the flow of moist, cold air into a region at low altitudes; or the transition from fog to cumulus.

Season: All year.

Range: Stratus may occur at any location in a region of humid, moving weather systems. Such conditions are met in the northern states and, in the winter, in the South. During very cold outbreaks in the northern regions of the United States, stratus may fill the arctic air mass for several days. Low-level moisture is always present or is refreshed each day by onshore winds, such as those along the coasts of the Gulf of Mexico or Pacific Ocean. Moisture also may be brought into a mountain valley and linger there for several days because no significant winds reach the low altitudes. As a result, the air reaches saturation under light-wind conditions during the evenings to form fog or stratus, which then burns off partially or completely each morning and early afternoon before starting the cycle over again.

Variations: The sky may be dark (opacus) and almost without any features as in plates 57 (nebulosus) and 58 (uniformis). When there is a visible feature, it may be a simple linear (undulatus) structure (56). Thin stratus (translucidus) may allow the visibility of the sun (59). Viewed against the side of a mountain (61), the base and top of a partial layer of stratus can be seen most often in morning or evening. A stratus layer (55, 60) near the ground may look like fog; the only real difference between fog and stratus is that stratus do not reach the ground, while fog reduces visibility at the surface.

Significance: An area of stratus represents saturation near the surface of the ground. Most likely, other clouds can be seen at the same location. Sometimes stratus are a recurring local condition along a coast or in an isolated mountain valley at certain times of the year. In other locations stratus accompany larger-scale, traveling weather systems, such as cold and warm fronts and tropical and subtropical weather disturbances. In the warmer months, stratus in the morning may represent areas of particularly high humidity that may become the location of cumulus development later in the day.

Comments: Stratus are generally diffuse and rather dull. These clouds are very common in regions such as coastlines and valleys. When larger-scale conditions are calm, stratus clouds form and clear in a cyclical pattern. At other times they may burn off quickly, introducing a day with only cumulus clouds in an otherwise blue sky.

62–67 Stratocumulus
Stratocumulus opacus, undulatus
Stratiform Clouds

Description: *Low-altitude, white to gray water-droplet clouds with a distinct cloud base and a variety of visible structures.* Generally, little precipitation falls from these clouds, except for a few very light and brief rain, snow, or sleet showers. Stratocumulus clouds are the lowest continuous stratiform clouds with noticeable variations seen from the ground. As a result, relatively small elements of the cloud may appear quite large from the ground. Stratocumulus may range from a few hundred feet above the surface (65) to a base at 6,500′ (1,981 m). Above that height similar-looking clouds are considered altocumulus; the variability in cloud appearance is less pronounced owing to their greater height.

Environment: Stratocumulus clouds indicate low-level moisture at temperatures ranging from somewhat above to well below freezing. For stratocumulus to form, adequate low-level moisture is needed so that saturation is reached. Relatively weak instability within a shallow layer of the cloud is also necessary to make stratocumulus clouds, which have a more visible structure than stratus clouds. Stratocumulus stop their upward growth within a few thousand feet or less above their cloud base and remain connected, while cumulus clouds grow more upward and become more separate. If the clouds reach a greater depth, they are considered cumulus. Often, a steady to strong wind helps produce the commonly observed long rows of stratocumulus (62, 63).

Season: All year.

Range: Stratocumulus occur throughout North America. In the northern states and Canada, they often occur during the warmer months after the wind has shifted to west or northwest and the atmosphere aloft is cooling. This situation often is present after the passage of a humid, traveling, large-scale disturbance, and is most visible during the warmest time of the day. In the wintertime, stratocumulus also are produced by such traveling systems in the southern states.

In the lee of the Great Lakes, out to about 100 miles (160 km) to the south, southeast, and east, and particularly in the winter months, the passage of cold and stable air across the warmer lakes often produces a moist, shallow layer of stratocumulus for several days, with a tendency for more clouds and snow showers during the daytime, when heating causes increased upward motion that causes condensation.

Along the coasts of the Gulf of Mexico and Pacific Ocean, low-level moisture is almost always present, or is refreshed each day by onshore winds.

Variations: For stratocumulus to be identified, the sky must have noticeable variations across it. Often there is an organization in the form of rows, lines, or patches (undulatus), as plates 62, 63, 66, and 67 show. The clouds may be thin enough to reveal the sun, or thick enough (62) to block it (opacus). A transition from fog and/or stratus is shown in plate 64. The very low cloud layer extending in a long row along the coast in plate 65 is probably the result of an interaction between the land and ocean.

Significance: Stratocumulus represent saturation and instability in a shallow layer near the surface of the earth. For this condition to occur, it is unlikely that many cumuliform clouds will be in the area at the same time, although stratocumulus may develop into cumulus when moisture increases and instability grows.

Stratocumulus may be a recurring local condition along a coast at certain times of the year. In other locations, stratocumulus accompany larger-scale, traveling weather systems, typically after the passage of a cold front or other disturbances, and especially in the afternoons. During the evening, fog or stratus are formed that partially or completely burn off each day during the morning to early afternoon. Stratocumulus are often seen before the sky clears, or may be the final result of the decrease in cloud cover before the cycle starts over again.

Comments: Stratocumulus are frequently organized in appearance and widespread over hundreds of miles. They occur along and downwind of some coastal and lakeshore regions, especially in the winter.

68–75 Nimbostratus
Nimbostratus nebulosus, opacus,
pannus, praecipitatio, virga
Stratiform Clouds

Description: *Dark gray to pale blue water-droplet*
precipitation clouds with noticeable blurring
in the area below cloud base.
Nimbostratus are rain and snow
clouds—their blurred look comes from
falling precipitation. In the summer
months, precipitation from convective
clouds is visible in only part of the sky,
and the full shape of other clouds,
which have a more complete outline or
appearance, can be seen. Blue sky may
also be seen in some directions.

Environment: Nimbostratus form either from
convection or from gradual uplifting
in a large, moving disturbance,
producing rain or snow. Lightning or
thunder in a nimbostratus cloud
changes its classification from a cloud
type to a thunderstorm.

Season: All year.

Range: Nimbostratus are visible at any location
where precipitation without lightning
occurs. Spotting nimbostratus in the
eastern and southern regions of the
United States is difficult, however,
especially in the warmer months when
haze and smog partially obscure the
sky. Thus, a nimbostratus cloud with
precipitation is hard to distinguish
from a diffuse-bottom cloud without
rain until it is nearly overhead.

Variations: There must always be visible
precipitation (praecipitatio) without
lightning (69, 71, 73) for the cloud to
be called a nimbostratus. In such a
region the sky can become quite dark
(opacus), as in plates 68 and 70.
Sometimes the rainfall will not reach
the surface (virga). The sky is often
without much form (68, 72, 75).
Shreds of cloud material (74),
sometimes called scud (pannus), may
occur below the primary cloud base

where saturation is reached, along the edge of the rain area.

Significance: Nimbostratus always bring precipitation. If a nimbostratus cloud is convective, a thunderstorm, with its associated phenomena, may develop rather quickly. Then, an assessment of the trend should be made over a period of minutes: Is the precipitating area increasing, moving, or dissipating? When precipitation moves toward a location, the associated threats of increased winds and higher waves (on water), the potential for lightning from a subsequent thunderstorm, and other effects, including heavy rain and flash floods in vulnerable locations, should be taken into account for the near future. Minor showers may occur quite often in the afternoon during the warmer months, though the impact may seem relatively minimal.

Comments: Nimbostratus, although common in some regions and seasons, are important clouds to watch; most thunderstorms start with a nimbostratus phase. All nimbostratus clouds should be monitored when planning outdoor activities.

CUMULUS CLOUDS

The presence of cumulus clouds is a sign of fine weather in the region. Shallow, vertical cumulus clouds are made when warm air rises, reaches its dew point, and then condenses. They appear light in color and have flat bases and rounded tops.

Cumulus become cumulus congestus when the upward currents of warm air continue. These clouds, if sighted early in the day, may signal the advance of stormy weather.

Cumulonimbus clouds (commonly called thunderclouds) yield precipitation and often bring thunderstorms. Gust fronts, occurring only in the presence of cumulonimbus clouds, can increase wind speeds to dangerous levels of over 100 miles per hour (160 kph).

Mammatus clouds are easily distinguished as well-defined, rounded "pouches" hanging from the underside of a cloud. They indicate that a thunderstorm is approaching or has passed. The base altitude for cumulus clouds ranges from about 2,000' (600 m) to 4,000' (1,200 m) in humid areas and as high as 10,000'–15,000' (3,000–4,500 m) in drier regions, with the cloud tops extending less than one cloud's width above.

76, 80–88   **Small Cumulus**
Cumulus humilis fractus, mediocris
Cumuliform Clouds

Description: Separate *light-colored water-droplet clouds,* with some *darker areas at cloud base.* Generally *flat bases and small rounded tops.* Tops extend less than one cloud width above cloud base (**80, 81**). Rain does not occur from these clouds because they are too shallow.

Environment: Small cumulus clouds are caused by

rising currents of warm air that cool to the dew point and condense. In humid areas, this condensation may cause cloud bases to occur at 2,000' (610 m) or below (86), or up to 4,000' (1,219 m) or somewhat higher (76). In drier regions, cloud bases may be as high as 10,000'–15,000' (3,048–4,572 m) (82). The humilis cloud top is limited because the atmosphere at the cloud top is too stable to permit growth. However, in the morning to early afternoon, the humilis may be only a temporary stage as some clouds continue to grow into congestus or cumulonimbus.

Season: All year.

Range: Humilis clouds with low cloud bases (87) are common in the eastern half of the United States, and along the west coast, at any time of the year. Higher bases are typical during the afternoon in summer in the western states (82), and during the afternoon in summer in the northern half of the United States and Canada during less humid periods.

Variations: When cumulus clouds are in areas with high wind speeds, they are torn into nearly horizontal fragments (82) with shallow depth (cumulus humilis fractus). When a cloud's size is approaching a depth that is greater than its width, it is a cumulus mediocris (81, 83, 88) in transition to the next stage, congestus. Humilis usually occur after a nearly overcast fog and/or stratus cloud cover has lifted and broken into cumulus clouds (84–87).

Significance: Presence of these clouds in the afternoon indicates no threat of showers in the near term. When cumulus humilis clouds are seen on a sunny day in the morning to early afternoon, the atmosphere is unstable enough for these clouds to grow into larger clouds. Humilis clouds are frequently present over most of the sky in the afternoon at the same time that large storms are also visible.

Comments: Small cumulus clouds are very common
in the United States and Canada during
the warm season, and at any time of the
year when there is an adequate supply
of low-level moisture and vertical
motion from larger-scale disturbances,
coastlines, or mountains and hills.
Humilis clouds sometimes grow into
larger cumuliform clouds.

77–79, 89–99 Swelling Cumulus
Cumulus Congestus
Cumuliform Clouds

Description: *Separate* or somewhat *organized white
water-droplet clouds* with some *darker
areas and rain often falling from them.*
Generally flat bases, and significantly
rounded tops with well-defined
outlines. Tops extend more than one
cloud width above cloud base (**91, 92,
94, 95**).

Environment: Rising currents of warm air have cooled
to the dew point, condensed into a
cumulus humilis, then continued to
grow into a deeper cloud. For a
cumulus to reach the congestus stage,
the atmosphere must be somewhat
unstable in a deeper layer than for
humilis. However, the congestus stage
may not be the last stage of growth,
since the cloud may continue growing
and become a cumulonimbus. Cloud
bases in humid areas range from below
2,000' (610 m) (**92**) up to 4,000'
(1,219 m) or higher (**90**). In drier
regions, cloud bases may be as high as
10,000' to 15,000' (3,048–4,572 m)
(**93, 89**).

Season: Primarily summer months.

Range: Congestus clouds with low cloud bases
are fairly common in the eastern half of
the United States, and less common
along the west coast, in the warmer
months of the year. In both of these
humid regions, adequate moisture is
often present during the summer to the

south of cold fronts and in the vicinity of large-scale upward motion caused by traveling disturbances. Congestus clouds are also formed frequently during the summer months within the relatively small zones of upward motion (less than about 25 miles/40 km) that are produced daily by sea breezes near oceans, lake breezes near the Great Lakes, and mountain breezes over large vertical changes in the land's surface. In drier climates, congestus clouds tend to occur with strong afternoon heating over or along the slopes of mountain ranges, during periods of moister-than-usual conditions, or in the vicinity of large-scale upper-level vertical motions.

Variations: In plate 89, the clouds are no longer flat enough to be humilis, and are in transition to congestus (note the rounded tops visible on many of the clouds). In plate 90, the congestus in the foreground is clearly deeper than the humilis in the background. In plates 91, 92, and 94, the clouds have tall vertical towers; all of them have some tilt because of different wind speeds at the tops of the clouds than at the bottoms.

Rain is falling from 78, 79, 93, and 98, but in all these clouds the tops are not fibrous or soft. In plate 78, the top of the cloud is precipitating into its lower part; since there is only rain and the cloud is not very tall, it is classified as congestus and not cumulonimbus. Clouds in plates 77, 93, 96, and 97 are well-developed and strong congestus at the transition to cumulonimbus. Since the clouds are large, organized, and beginning to have a softer appearance at the top (typically indicative of ice), these clouds probably did not stop growing at the stage shown in these plates. In plate 99, the congestus clouds are aligned in rows, which appear like walls from the side. Such "lake-effect" clouds form during the colder months on the downwind side of the Great Lakes, occasionally producing very heavy snow. They are rather shallow (typically less than 10,000′– 15,000′/ 3,048–4,572 m), rarely produce lightning, and show some rounded cumuliform tops. Since they are partly diffuse on top, they can be classified as congestus or cumulonimbus, but because of their shallow depth and lack of any anvil shape on top, they are probably congestus.

Significance: Precipitation may fall from congestus clouds as they grow deeper. In the afternoon, their presence indicates some threat of showers in the vicinity. When seen on a sunny day in the morning to early afternoon, the atmosphere is probably unstable enough for them to grow into cumulonimbus. If these are the largest cumuliform clouds in the region by late afternoon, they most often represent the end of the growth process for cumulus on that day, unless a larger-scale disturbance, or other factors not apparent from the sky alone, will help them grow further. Congestus clouds may exist in one part of the sky in the afternoon simultaneously with large storms and humilis clouds in other directions. In the tropics, including Florida in summer, cumulus clouds often reach only the congestus stage with very heavy rain in the form of showers for a portion of an hour, and occasionally in lines that may last for several hours.

Comments: A somewhat common cloud in the warm season over all North America whenever there is a good supply of low-level moisture and some additional forcing due to other factors. The earlier in the day that congestus clouds form, the more significant they are as indicators of strong storms later that day or in the region.

100–127, 129, Cumulonimbus
130 Cumulonimbus calvus, capillatus,
incus, pileus, spissatus
Cumuliform Clouds

Description: *Isolated* to *highly organized clouds made up
of water droplets in lower portions*, and *ice
particles in upper portions*, with *dark bases
and with precipitation falling from them*.
Generally flat bases unless rain is
falling. Tops on much of the
cumulonimbus are diffuse and soft,
blowing downwind, and without many
sharply defined outlines.
Cumulonimbus clouds are deep enough
that *cloud bases* (112) *often are not visible
at a distance when the rest of the cloud can
be seen* (100, 105, 111, 114).
Precipitation always falls from
cumulonimbus clouds, although, in
dry regions, it may evaporate before
reaching the ground (102, 103). Such
*evaporating precipitation is known as
virga.* Severe weather (strong winds,
heavy rain, etc.) occurs with
cumulonimbus clouds.

Environment: Rising currents of warm air have cooled
to the dew point, condensed into a
cumulus humilis, then a deeper
cumulus congestus, and finally a
cumulonimbus. For the cumulus to
reach the cumulonimbus stage, the
atmosphere must be quite unstable in a
deep layer. Cloud bases in humid areas
range from below 2,000' (610 m) (112)
up to 4,000' (1,219 m) or higher
(111). In drier regions, cloud bases may
be as high as 10,000'–15,000'
(3,048–4,572 m) (101, 103).

Season: Warm months throughout the
United States.

Range: Cumulonimbus clouds with low bases
occur in the eastern half of the United
States in the warmer months of the
year, but are quite rare in the humid
regions of the west coast because of dry,
subsiding air aloft. In humid areas,
adequate moisture and instability aloft
are needed during the summer to the

south of cold fronts and in the vicinity of large-scale upper-level vertical motion. In drier climates, cumulonimbus clouds tend to occur on days with strong afternoon heating over or along the slopes of mountain ranges; during periods with very moist conditions aloft; or in the vicinity of strong, large-scale upper-level vertical motion.

Variations: The cumulonimbus clouds in plates 100, 101, 104–111, 113, and 114 are in transition from congestus to cumulonimbus, showing tops with soft and diffuse features (cumulonimbus calvus) instead of sharply defined outlines. The two sequences in plates 104–107 and plates 108–110 show the transition at the top of the cloud from the relatively clearly defined edges of congestus to the ice phase of a cumulonimbus. Plates 104–107 also illustrate a horizontal pileus cloud draping across the cumulus beneath, because of a relatively shallow layer of air near saturation being lifted by a strong updraft, then finally merging with the main cumulus as the cumulonimbus stage is reached. (Most updrafts that are strong enough to produce pileus will result in a cumulonimbus.)

Several clouds (100, 111, 114, 116) show a separate horizontal layer near the top, where the cloud hesitated in its vertical motion because of less moisture or vertical instability, formed an anvil-like horizontal cloud layer, and then resumed its upward motion, as occasionally shown by a rounded bubble on top of the anvil for a short time (111). More classic anvil (incus) shapes are shown in plates 119, 122, 124, 126, and 127. A cloud with an anvil is always a cumulonimbus, since it shows the cumulus to have stopped growing at a high level, usually above 30,000′ (9,144 m), because of strong stability (warming or drying) above the anvil's top. Anvils may blow off for a long distance (120, 121) and become detached or much larger than the parent cumulus cloud that generated the anvil. Sometimes, the anvil may partially detach (113), or be the only part of the cumulonimbus that remains (118, 123).

A full cumulonimbus and anvil structure (129) may also be formed in a cold winter environment. Some anvil tops will blow downstream in a cirriform shape called capillatus (127, 129). Cirrus spissatus refer to anvils thick enough (125–127) to be gray on the side away from the sun. As cumulonimbus clouds grow beyond the simpler single-cloud structures shown in these plates, and become arranged in rows or complexes or have large areas of updraft bubbles (112, 115, 117, 119, 130), their association with severe weather becomes much more widespread.

Significance: Precipitation always falls from cumulonimbus, but sometimes it evaporates partially or completely in areas where the lower levels of the atmosphere are very dry. Severe weather accompanying cumulonimbus can include tornadoes, waterspouts, and funnel clouds; brief to prolonged heavy rain, hail, sleet, or snow, and accompanying flash floods; strong winds and turbulence at any level of the atmosphere, including microbursts; and lightning within clouds, between clouds, and from cloud to ground. When cumulonimbus clouds are seen on a sunny day in the morning to early afternoon, the atmosphere is probably unstable enough for further cumulonimbus development into more intense and larger organized systems with accompanying significant to severe weather in the region. If these are the largest cumuliform clouds in the region by late afternoon, they most often represent the end of the cumulus growth process on that day, unless a larger-scale disturbance or other factor (not apparent from the sky alone) will help them organize and intensify further toward evening.

Comments: Called a "Cb" in the aviation and meteorological communities (and abbreviated as such on a weather map), cumulonimbus are very important clouds to track whenever there is a good supply of low-level moisture and strong additional forcing due to other factors. The earlier in the day that a cumulonimbus forms, and the more organized or vigorous it appears, the more significant it is as an indicator of a wide variety of important storms and accompanying weather later that day or elsewhere in the region.

132–135, 194, 195 **Mammatus**
Cumulus mammatus
Cumuliform Clouds

Description: A series of *pouch-shaped cloud elements hanging downward from a middle or upper cloud layer.* These are always part of the underside of the anvil blowoff from a cumulonimbus. *Gray to pale blue clouds* that vary in size and dimension.

Environment: Mammatus clouds indicate a very moist and unstable middle or upper level in the atmosphere, overlying a drier layer below an anvil produced by an adjacent cumulonimbus. Sometimes the cumulonimbus may be too far away for the updraft to be visible, but there always has been one in the vicinity at some earlier time. The individual elements of mammatus represent convection in the inverse, that is, instability in the downward direction rather than upward. This situation exists because of the large amount of moisture and heat flowing out from a cumulonimbus anvil over the neighboring, relatively undisturbed air.

Season: Warm months in eastern United States, and in dry and/or mountainous areas.

Range: Mammatus clouds occur only in the presence of cumulonimbus. This situation can exist in the eastern states during warm months, rarely in the humid regions along the west coast, and frequently on summer days in drier and/or mountainous regions.

Variations: Spectacular mammatus clouds may be large and cover most of the sky (133, 135, 194). The more common view (132, 134, 195) shows weaker variations over a part of the sky. Sometimes they appear to radiate outward from the cumulonimbus updraft (132).

Significance: The presence of mammatus clouds often indicates a vigorous cumulonimbus in the vicinity. Mammatus clouds more often follow the most active growth stage. The time interval ranges from a few minutes, in the immediate vicinity of a small thunderstorm, up to an hour or more, when the mammatus may drift over a location up to 25 miles or more away from the parent thunderstorm. However, a sky mostly filled with large mammatus clouds indicates a very strong thunderstorm was or is nearby, or may be approaching.

Comments: If the mammatus clouds are relatively weak, in the distance, and not approaching, the threat from the storm that produced them is not likely to be very great. However, if they are large and approaching, the sky should be watched closely for the possibility of severe weather.

128, 131, 196, **Gust Front**
197, 199 Cumulus arcus
Cumuliform Clouds

Description: *A long, horizontal, arc-shaped low cloud produced by a gust front.* The gust front is *the leading outflow of cooler air from the base of a cumulonimbus cloud,* so it occurs on the outer edges of these lower clouds. The gust front may have an anvil of cirrus clouds overlying it at higher levels and flowing in the opposite direction.

Environment: Gust fronts form when cooler air, accompanied by rain during the downdraft stage of a thunderstorm, begins to move away from the parent thunderstorm's rainfall area. When the outflow's air is much cooler than the surrounding air, when there is enough low-level moisture in the region, and when several other complex features exist in the wind and temperature structure aloft, the outflow can become highly organized and form a cloud boundary that may extend for up to 100 miles or more (161 km) in a continuous, curved line, marked by low cumulus or stratus clouds.

Season: Warm months in the eastern United States; sometimes in summer in dry or mountainous areas.

Range: Gust fronts occur only in the presence of cumulonimbus clouds, and are visible near relatively few thunderstorms. Storms with the potential for gust fronts occur on many days in the eastern states in the warmer months of the year, rarely in the humid regions along the west coast, and sometimes on summer days in drier and/or mountainous climates.

Variations: Gust fronts occur more often in partial semicircular shapes (131) than in long, organized lines. When very well organized gust fronts (128, 196, 197, 199) occur, they may cover most of the sky. When some thunderstorms move off mountains during the afternoon or evening, especially in more arid regions where the evaporation of rain makes the cooling of the downdraft stronger, they are able to produce relatively strong gust fronts, similar to the ones in plate 128. These gust fronts might not otherwise have been as strong, since the downdraft is intensified by the sloping terrain.

Significance: A gust front of any size or shape should be regarded with caution, since there is a shift in the surface wind speed and direction when the line passes overhead; this is especially hazardous to aviation interests. Sometimes, the wind may not change much during the first few minutes after a gust front passes, then the speed may steadily increase over the next 15 minutes. Wind speeds greater than 100 mph (161 kph) occur several times a year in the United States over an area the size of several counties in association with gust fronts that have become highly organized into squall lines.

Comments: If a gust front is approaching, there will probably be a change in wind direction, and always an increase in wind speed for at least a few minutes. Sometimes, a large moving thunderstorm may first be felt in a given location by the arrival of its gust front. Then an increase in rain may follow, with frequent lightning while the gust front passes overhead. The gust front, then, should be watched both as an important feature in itself, and as an indicator of possible future significant or severe weather.

OROGRAPHIC CLOUDS

The lifting of air caused by the slope of a hill or mountain creates orographic clouds. An air current cools as it rises up a slope and, reaching its dew point or condensation level, forms a cloud. If the cloud is at ground level, it is fog. Orographic clouds can be of either the horizontal stratiform type or the vertical cumuliform type. Both indicate areas of high winds.

138–141 Mountain-induced Cumulus
Orographic
Orographic Cumulus Clouds

Description: *Cumuliform water-droplet clouds with rounded tops that remain relatively stationary over a terrain that has a major change in elevation.* A ground feature that rises several hundred or more feet is usually near the cloud formation, although these clouds may occur as far as several miles downstream from a hill or mountain. When viewed through time from one location, orographic cumulus clouds appear stationary, although individual cumulus clouds form, grow, and dissipate downwind from the forcing feature over a period of 1–20 minutes.

Environment: For orographic cumulus clouds to form, the presence of adequate moisture and vertical instability is necessary. The moisture is condensed into water droplets by the forcing terrain, and the vertical instability prompts convection.

Season: Primarily in summer months.

Range: Orographic cumulus clouds sometimes occur over the hill and mountain ranges of the eastern United States and over the mountains of the West. Stationary cumulus clouds generally require individual terrain features at least 500′ (150 m) high in humid regions, and still higher in more arid areas.

Variations: Depending on the height of the
moisture layer, clouds may form as
wind flows up a mountain (139), near
the top in the form of a banner cloud
extending downwind (140), at the top
(138), or higher than mountaintop level
(141). Plate 141 shows a combination
of both cumulus and stratiform
orographic clouds. Occasionally a wall
of clouds with some imbedded cumulus
shapes may develop over a mountain
ridge, representing the presence
of moisture along the top of the
mountain or hill; downslope flow
then results in warming at lower
elevations downstream.

Significance: The high winds of orographic cumulus
clouds mainly impact their immediate
area. They indicate where subsequent
thunderstorms may grow later in the
day if wind flow, stability, and
moisture conditions remain constant.
However, mountains induce
complicated diurnal wind-flow patterns
that change throughout the day, so that
orographic cumulus in one area at one
time will not necessarily remain there.
Flying aircraft are frequently affected
by the rather intense horizontal and
vertical motions within and near
these clouds.

Comments: Cumulus orographic clouds represent
areas of high winds in the local area and
sometimes on the ground beneath
them. The presence of these clouds
should raise caution for aviation-related
activities.

136, 137, **Mountain-induced Stratiform**
142–149 Orographic lenticularis
Orographic Stratiform Clouds

Description: *Stratiform water-droplet clouds with very
smooth edges and features* that remain
relatively stationary over or near terrain
that has a major change in elevation.
These stratiform clouds occur in a

region with a surface-terrain feature, although the clouds may be as far as 50 miles (80 km) away. Orographic stratiform clouds, generally downstream from a hill or mountain, appear stationary; however, portions of the clouds form, grow, and dissipate within several hours.

Environment: For orographic stratiform clouds to form, adequate moisture must be present at optimal altitudes (relative to the terrain) so the moisture can be condensed into water droplets by the upward motions due to the mountain or hill by relatively strong winds blowing perpendicular to the mountain barrier. Because the airflow is usually quite strong, very clear skies typically accompany these clouds.

Season: Primarily in winter months.

Range: Orographic stratiform clouds usually occur in the United States over the western mountains, and occasionally over the hills and mountains of the eastern United States. Orographic stratiform clouds generally require individual terrain features at least 500' (150 m) high in humid regions, and still higher in more arid regions.

Variations: How the stratiform orographic cloud looks depends on the height of the moisture layer and the vertical variations in the speed and direction of the wind flowing through the cloud. Clouds may form as wind flows up the mountain and down the other side in a cap, or crest (145), at the top (143), higher than mountaintop level, or much higher yet in the lenticularis (lens) shape (142). Delicate variations in thin layers of moisture aloft were made visible by lift from the mountains in plates 136 and 149. Multiple layers in plate 146 are capped by a pileus cloud. The Kelvin-Helmholtz waves in plate 147 are forming full circles as they move downwind; they represent very strong wind shear between the bases and tops of the clouds. Sometimes the resulting cloud structures are simpler than the underlying complex topography (144, 148). But sometimes the clouds are more complex than the terrain, as in plate 137, where exaggerated updrafts on the left sides of the two cloud elements turn sharply into horizontal flow. Plate 141 shows a cumulus formation that has developed a stratiform veil on top. Sometimes a stratiform cloud over or downwind of mountains will produce excellent iridescence (see plates 333 and 334).

Significance: Orographic stratiform clouds primarily affect their immediate area. They indicate where strong wind flow and associated turbulence is present. When the wind flow is perpendicular to the barrier, it often warms as it flows downhill, as in the westerly chinook of the Rocky Mountains or the Santa Ana winds of southern California. These chinooks occur mainly over and close to the mountain ranges; speeds may reach up to 150 mph (241 kph), and the winds may maintain their strength for quite a few hours, most often at night. When the wind blows more from the north, the air stays cold, as during the northwesterly bora of the Rocky Mountains. The bora's speeds of up to 100 mph (160 kph) may spread onto the lower downwind plains out to about 50 miles (80 km); boras have less preference for nighttime hours. These patterns change continually as the weather systems causing them pass overhead, so that orographic stratiform clouds in one area will not necessarily remain in that area. Aviation-related activities are very hazardous due to the intense horizontal and vertical motions within and near these cloud forms.

Comments: Stratiform orographic clouds represent areas of high winds aloft and sometimes on the ground beneath them.

MIXED SKIES

The sky is in a constant state of flux, exhibiting many stages of cloud growth and development. To observe clouds and decipher their messages regarding impending weather, you must watch the sky frequently over the course of the day. Several cloud types can coexist at different levels of the atmosphere, with wind speed, temperature, and moisture content varying from layer to layer.

The mixture of clouds in the sky, and each one's effect upon another as they transform, dissipate, and blend, in turn determine what kind of weather we receive.

150 Altocumulus and Cirrus Clouds

Description: This plate shows the somewhat irregular patterns of two cloud types, altocumulus and cirrus. Here, sunset illuminates and colors the clouds, making the different features of this chaotic sky more clearly visible. While the lower altocumulus clouds are turning yellow, the cirrus clouds are high enough to receive direct sunlight and thus stay white for a while longer. The swirl in the upper right corner is distinctive of cirrus clouds.

Environment: Moisture adequate to form a partial cloud cover is probably available in both the middle and upper levels of the sky. However, it is not an especially significant situation, or more clouds would have formed.

Significance: Daytime heating will probably not cause much more cloud cover unless there is a large-scale disturbance traveling into the region.

Comments: Sunset and sunrise are excellent times for distinguishing layers of cloud cover. When the sun is low on the horizon, it illuminates and colors the various cloud layers from such an angle as to reveal most readily their individual character and relative positions in the sky.

151 Cirrus and Cumulus Clouds

Description: The clear distinctions between cirrus and cumulus shapes are easily seen here. Diffuse cirrus fibratus clouds cover the upper portion of the view, while rounded shapes with flatter cumulus bases appear below.

Environment: Adequate moisture exists at the upper level of the atmosphere, allowing for the formation of cirrus clouds there, while vertical motions due to low-level instability, combined with a source of moisture in lower levels, have formed the cumulus humilis. Neither formation process is very widespread or prolonged.

Significance: This relatively benign situation occurs commonly during fair weather, and in such instances neither cloud type is very large or intense. When cumulus with intense updrafts and/or thick cirrus clouds move into a nearly clear sky, the weather situation is no longer steady.

152 Cumulus and Altocumulus Clouds

Description: Tall, vertical cumulus congestus towers are accompanied by horizontally arrayed altocumulus. The bases of the cumulus clouds are quite low to the ground, and thereby blend into the lower altocumulus. Such formations may lead to showers and thunderstorms that will occur in the vicinity, and some could produce heavy rain due to the deep moisture layer that is available.

Environment: This photo was taken in Florida during the summer, when a significant amount of moisture is present in the lower levels of the atmosphere. Under these conditions, the cumulus congestus clouds grow vertically with no restrictions. The bands of altocumulus visible in the lower portion of the image have formed due to the presence of moderately large amounts of moisture in the atmosphere.

Significance: The cumulus congestus clouds are fairly common for this region and time of year. However, the coexistence of midlevel moisture evidenced by the middle clouds indicates a more active situation than normal.

153 Cumulonimbus, Cumulus, and Altocumulus Clouds

Description: This photograph, taken near Jackson, Wyoming, shows several weak cumulonimbus clouds in the mature to dissipation stage with virga falling from them, and cumulus congestus in the growth stage. No significant growth of other cumulus clouds appears to be taking place at this time. A patch of altocumulus, visible in the upper portion of the image, has been left behind by one of the earlier cumulonimbus.

Environment: The cumulonimbus clouds probably formed and grew over the high mountains surrounding the valley, then were moved by the prevailing wind onto the lower elevations, where they stopped growing (as suggested by the lack of any well-defined flat bases). Adequate moisture for further growth was available over the mountains but not elsewhere, resulting in the clouds' dissipation.

Significance: New cumulus clouds can be seen forming over the valley as the afternoon wears on, indicating some instability for upward motion in the air mass. The clouds appear to be struggling, however, and will not reach the height of the cumulonimbus that grew over the mountains earlier in the day. Very light showers from any virga reaching the ground may occur. Some gusty winds around the virga may also occur, but would probably not be a threat.

Comments: This view is typical of many summer days in the western United States and Canada. The mountains provide the heating and updrafts that result in forcing, allowing cumulonimbus clouds to grow for a short time until they are blown away and die over the valleys. Without the mountains, the cumulus would not reach beyond the congestus stage.

154 Cumulus, Cumulonimbus, Altocumulus, and Cirrus Clouds

Description: Taken at a high altitude in Colorado in early summer, this plate shows a large, icy cumulonimbus cloud precipitating, with other cumulus congestus clouds growing along the mountain ridge. The cumulonimbus has the characteristic incus, or anvil, shape, but the lower virga appears to be an ice-crystal form without the more typical rainshower appearance. New development of cumulus congestus is evident along the Continental Divide across the entire view. An altocumulus patch at upper left is probably the remains of an earlier cumulus cloud that died. A few cirrus wisps appear at right.

Environment: Cumulonimbus clouds are the result of vertical instability. One of the several ways they can form is by a deep moist layer starting vertical motion from a low cloud base. In this case, the thunderstorm probably grew because of surface heating of the elevated land mass combined with rather cold temperatures aloft, conditions that created a large temperature difference in the vertical air during the afternoon.

Significance: Several of the cumulonimbus that formed earlier on this day have reached their maturity. Some of the new clouds along the Divide may later reach cumulonimbus stage, but their growth may be inhibited by the cooler air at the ground that was caused by the outdraft from the earlier storms. Such thunderstorms at this altitude and season often produce isolated moderate rain, hail, sleet, and winds flowing outward at the ground away from the rainshafts.

Comments: These clouds are characteristic of the mountains in the western United States. The variation apparent here— with clouds made up primarily of ice crystals rather than water droplets— may also occur during cooler months outside of summer.

155 Cumulus, Altocumulus, and Cirrus Clouds

Description: A variety of clouds from low, middle, and high altitudes are distinguishable here. The cumulus above the smallest cumulus humilis are somewhat developed but have only ragged cloud bases (fractus), and their tops are diffuse and disorganized. The middle altocumulus clouds, at upper right, are scattered and not very thick; at far right, high-altitude cirrus clouds are also scattered across the sky.

Environment: Sufficient moisture is present to cause cloud cover at several layers of the atmosphere. It is somewhat unusual to see this much cloudiness at all three altitude levels without at least a few clouds being better developed.

Significance: The existence of clouds at so many layers may be indicative of a larger-scale weather system moving through the region. Watch this type of sky for any major cumulus or layered cloud development.

156 Altocumulus, Altostratus, and Cirrostratus Clouds

Description: These highly organized, multilayered clouds are intense toward the horizon because of the moist airflow associated with the southwest United States monsoon. Nearly overhead is a series of altocumulus clouds that merge at lower left into altostratus bands. Above the altostratus is a layer of cirrostratus. The clouds directly beneath the blue sky are white to light gray, while clouds under the cirrostratus are quite dark because the light cannot penetrate the intervening cloud layers.

Environment: The southwest monsoon is an important weather factor from July to September in the Four Corner states (Utah, Colorado, New Mexico, and Arizona), and sometimes in areas to their north and east. This photograph, from Colorado, was taken as a plume of moisture was passing nearby. While the mountains to the west have removed the very humid air near the ground that produced thunderstorms to the southwest, the upper-level moisture can be seen streaming over them in the form of multilayered clouds.

Significance: A significant weather system is often marked by the presence of several nearly continuous layers of clouds that are stacked one atop another and occur in bands. At times no convection will take place, as is the case here, while at other times the cloud layers will be accompanied by cumulonimbus, as in plate 165. When the monsoon looks this way in the morning, the potential for flash floods and other severe weather is great, because cloud development can be rapid and widespread once daytime heating begins.

Comments: Highly organized layered clouds like these, with or without cumuliform clouds in their makeup, appear only in the presence of a weather system that extends through deep layers of the atmosphere. Each summer during the southwest U.S. monsoon, these conditions exist on few days in any single location, and their presence indicates a potential for severe weather. It is important to note their occurrence and take precautions when planning outdoor activities, especially around rugged terrain.

157   Orographic Cumulus and Cirrus Clouds

Description: Lines of clouds are seen crossing each other at nearly right angles in a mountainous region of Wyoming. Not far above the mountain ridge, two parallel lines of cumuliform clouds, orographic cumulus probably in a north-south direction, have resulted from westerly-component winds flowing across the ridge. Air rises and sinks in a wavelike form, visible at the top of the updrafts. Above this pair of lines is a series of roughly linear cirrus elements that are flowing along with the westerly wind at upper levels. Some of them are cirrus spissatus, and probably were anvils from cumulonimbus clouds that were in the sky earlier in the day.

Environment: Orographic clouds require that a portion of the horizontal flow be converted into upward motion, so the wind flow is likely to be from the west at a moderately strong speed at upper levels. The possible anvils are being blown far away enough to be dissociated from the original clouds. Not much moisture exists, or the cumulus would be better developed.

Significance: The rows of orographic cumulus clouds indicate significant wind flow in the immediate region and the chance of increasing wind speed.

158 Cumulus and Cirrus Clouds

Description: A remarkable patch of cirrocumulus clouds dominates the sky, merging with an area of cirrus near the horizon (although it is difficult to tell whether these clouds are continuous or at different levels). In the upper part of the image, the individual elements of the cirrocumulus appear somewhat larger than usual, but their small size in the same cloud layer to the right shows the type clearly. Some cumulus humilis are scattered at lower levels about 2,000′ (600 m) above the ground.

Environment: The presence of cirrus clouds over a large region indicates a reasonably high amount of moisture at the cloud level. The cumulus, however, show only weak vertical development, meaning that the moisture does not extend very far down from the cirrocumulus.

Significance: On warm to hot days in the summer over most regions of the United States, such a large upper-level cloud layer indicates that significant cloud development could occur later that day or evening. This image, however, offers no other visible clues that might signify a threat of greater cloud development. Nonetheless, by late afternoon or evening, it is possible that a traveling squall line or other disturbance may use the moisture at the cirrus level to grow into a larger cloud system.

159 Cirrus Clouds

Description: Several different types of cirrus are visible simultaneously. Streamers of virga can be seen falling from individual cirrus floccus clouds in the upper part of the image. A broader expanse of cirrostratus appears on the horizon over the plains of Kansas.

Environment: There is not enough moisture in the lower layers of the atmosphere to support any cumuliform cloudiness. At higher levels, however, small cirrus clouds form in flocks and nearly saturated conditions exist, so that ice crystals can fall for quite a distance before disappearing completely.

Significance: Moist conditions at upper levels are indicated by the cirrus streamers and the larger region of cirrostratus near the horizon. The cirrostratus are possibly due to outflow from the tops of cumulonimbus, although the original storms are not visible at this distance.

Comments: The cirrostratus in the distance may be a nearly continuous layer, and may actually be the tops of thunderstorms in the distance. However, the sky appears very clear here thanks to the visual compression of clouds near the horizon. We look through much more of the atmosphere near the horizon than overhead, which makes it appear that there are more clouds along the horizon than is actually the case. Watch the clouds just above the horizon for as long as possible when trying to judge whether or not distant clouds are truly more widespread or increasing.

160   Cumulonimbus, Stratocumulus, and Altocumulus Clouds

Description: This sky is deceptively disorganized and unimpressive, but a thunderstorm is visible through the moderately pervasive haze that makes all the clouds appear to be shades of light blue. The outline of the top of an active thunderstorm can be seen at the upper-right corner of the image. This cloud is large and opaque, with a somewhat rounded bubble that is most likely the top of a cumulonimbus that is beginning to produce a cirrus cloud in the incus phase. (Compare this cloud to plates 100, 101, 104–11, 113, 114, and 123, where cumulonimbus with rounded tops are in transition to anvil shapes.) Here, the sun casts a shadow, from upper right to lower left, onto a lower altocumulus cloud. The other clue to a storm in the area is the clearly defined stratocumulus line along the horizon that probably represents a gust-front boundary. The altocumulus here has a more windblown appearance compared to common altocumulus that has rows, lines, and more repetitive patterns.

Environment: Abundant moisture is present, enough to create a hazy lower atmospheric layer, a widespread altocumulus layer, and a thunderstorm. Winds appear to be relatively light, due to the lack of any lines, rows, or other patterns in any of the clouds except that imposed by the outflow circulation induced by the cumulonimbus.

Significance: A thunderstorm that has grown in a hazy environment has no fewer of the attendant hazards than one that has formed in better visibility, and should be considered an equal threat for wind, lightning, hail, high winds, or heavy rain. The haze simply masks the visual clues that warn of a coming storm. Sometimes, the first hint of a cumulonimbus in the area is close thunder; then precautions must be taken quickly.

Comments: Although thick haze may hide clouds in some areas for long periods, particularly in the eastern and southern states in the warm months, clouds of significance can be present. Knowing what to look for in the smaller areas of sky, where cloud shapes may be discernible, often makes it possible to anticipate storms in sufficient time to avert danger.

161 Altocumulus, Cirrus, and Orographic Clouds

Description: The sky is filled with orographic clouds in a wide variety of sources, shapes, and altitudes. In the distance are the Rocky Mountains, aligned on a north–south axis. A deep layer of clouds is apparent across the entire horizon, parallel to the mountains and above and behind them; altostratus occur directly above the ridge, and a deep cirrus layer occurs above that. At the very top of the image is an altocumulus layer with the same north–south orientation as the distant mountains. Below that is another elongated line of altocumulus, as well as several more strands farther away toward the mountains. The only nonorographic clouds seen here are small wisps of cirrus in the lower portion of the view.

Environment: The atmosphere is filled with winds, perpendicular to the mountain range in a relatively deep layer, that are strong enough to cause downstream clouds from the mountains more than 25 miles (40 km) away. Although it is not possible to tell what the wind speeds are from the visual clues alone, generally a sky full of organized clouds with such alignment as one sees here requires wind speeds in excess of 50 mph (80 kph). A moderate amount of available moisture makes the clouds visible, especially over the mountains, where the cloud layer is quite deep.

Significance: Such clouds are fairly common outside the summer months in the lee of the Rocky Mountains from Canada to New Mexico. They do, however, represent significant to extreme conditions of horizontal and vertical wind shear that can affect any aviation in the region. The orographic winds accompanying these kinds of clouds can also have strong effects on ground activities. Two features that make such situations unique in the United States and Canada are their tendency to last many hours (up to a day or more) and to increase in speed at night.

162 Cumulonimbus, Cumulus, Altocumulus, and Cirrus Clouds

Description: This view is typical of the way a thunderstorm appears when it is in the mature-to-dissipating stage and the sun is behind it. Here, the tops of a large cumulonimbus appear generally soft as they are outlined by the sun shining from behind in a moist environment. Around the base of the larger cumulonimbus are smaller cumulus congestus. Also apparent are some patches of altocumulus in the sunlight at center, and in the shade at upper left. A thin layer of cirrostratus is visible above the cumulonimbus, thickening toward the top of the view.

Environment: A large amount of moisture must be available in several layers for a sky like this to develop. This photo was taken in Arizona during July when the southwest monsoon was very active. Moisture is flowing from the Gulfs of California and Mexico in large, organized areas that are nearly saturated through much of the atmosphere. Cumulus activity begins early on such days, and usually reaches maturity by the end of the day. The other cloud layers also indicate where nearly saturated conditions are present.

Significance: This thunderstorm is in the mature phase and will not likely grow much larger. Moderate gusts of wind may occur some distance away from the parent cloud. In desert regions such as this, similar clouds may produce brief, heavy rainfall that can rush down dry washes and produce flash floods in low-lying roads and other areas.

Comments: Multiple layers of clouds with embedded cumulonimbus always indicate the presence of a large-scale disturbance. Precipitation may be quite heavy in some regions, usually occurring from afternoon to nighttime.

163 Cumulus and Cirrus Clouds

Description: Dramatic streamers of cirrocumulus clouds with delicate internal variations overlie a benign field of cumulus congestus. In a disconnected layer at the bottom of the atmosphere are afternoon cumuliform clouds, with their typical rounded tops and flattened bases. They are not able to break through a dry and/or warm layer to develop into cumulonimbus clouds.

Environment: Sufficient moisture is available at high levels to produce well-developed cirrus clouds, and surface heating has produced numerous cumulus clouds. However, there is no connection between the two cloud-bearing layers, because the air between them is inhospitable to any cloud development.

Significance: This type of cloud view is common during warmer months in the United States and Canada, although the upper clouds are more fully developed than usual. This may indicate the potential for more moisture flow, which would cause further cloud development later in the day.

164 Cumulus, Altocumulus, and Cirrus Clouds

Description: A series of cumulus congestus clouds is producing rain in the very moist, oceanic region along the coast of Florida. Above them is a nearly transparent layer of cirrostratus clouds. Patches of altocumulus at right are probably remnant water-droplet clouds from other cumuliform clouds that have dissipated. Of particular interest is the row of tall cumulus aligned from left to right, probably in the direction of the prevailing wind, or the shear of the wind in the vertical. The cloud at left still has a hard base; the cloud at center has a well-defined rain shaft; and the areas to the right may be partial remnants of earlier showers. The tops of these clouds are still sharply defined in their outline, and not diffuse or streaky in their windblown upper portions, so that it is more appropriate to call them congestus than cumulonimbus. Some small bubbles on top remain when they are no longer connected to the lower parts of the clouds. This cloud view is quite typical of the subtropics in the summer, and tropical regions all year.

Environment: The behavior of the cumulus clouds indicates the condition of their environment. At the base of the cloud at left, there is sufficient moisture and heating to initiate cloud development over the warm ocean; the updrafts reach upward from the base in just one rounded bubble, representing an updraft pulse, and not in an organized group of updrafts and bubbles. When the updrafts reach a height where the air is drier and/or warmer than below, the updrafts level off, as the clouds at left and center have. A short, quick growth spurt in a moister and/or colder layer above that level sustains a very small cumuliform tuft.

Significance: This cloud sequence is so distinct that no greater growth from cumulus clouds is expected to occur in the region. Any new clouds that form will produce brief, relatively heavy rainshowers. The atmosphere appears to be unable to make a cumulonimbus at this time and place, due to a combination of temperature stability in the vertical, low humidity, and weak wind shear. The cumulus congestus also will result in some brief gusts of wind in the vicinity of the showers.

165 Stratocumulus, Altocumulus, Altostratus, and Cirrostratus Clouds

Description: Widespread clouds in multiple shapes and at different atmosphere levels appear increasingly intense and dark toward the left, indicating an important large-scale disturbance in the vicinity. This photo was taken in the Florida Keys in late fall. In the bright area at the upper right, sunlight shines through a thin cirrostratus layer that is probably above all the other cloud layers. Across the ocean is a long line of stratocumulus clouds organized by the disturbance in a location that is not related to the land-sea boundary, because the daily heating cycle is being overwhelmed. Toward the left is a very dark, somewhat vertical cloud region that appears to be precipitating from the middle levels, and may be remnants from a distant or preceding cumulonimbus. Between the brightest and darkest regions are several layers and patches of altocumulus and altostratus.

Environment: Although a surface cold front passed through the region earlier, the most important feature of this view is the flow of moisture from the southwest in a deep layer at upper levels. This flow, which also was moving upward over a broad scale, brought in clouds from the subtropics, resulting in a deep, saturated layer. When this situation comes to pass during the fall and spring months in the southern states—especially in Florida—it is not the surface frontal positions that most readily reveal where the important rainfall and thunderstorms will occur, but the large-scale upper-atmospheric flow that is the dominant factor.

Significance: Sometimes a sky such as this one—full of clouds with rain falling from a few of them—will occur in humid regions during the summer at the end of a day and result in very active thunderstorms that may continue into the night. On the next day, widespread clouds and rainfall may reappear much earlier in the day than usual because of lingering upper-level moisture. The counterpart to this humid-region sky in the drier monsoon region is shown in plate 156.

166 Cumulus Humilis, Cumulonimbus, and Cirrus Clouds

Description: Note the two conspicuous, detached cumulonimbus anvils with small showers falling from them. Their existence is a clear indication of instability, yet their small and isolated condition—there is no thick middle-level cloud layer—shows that this instability is not strong enough to produce more than a few weak cumulonimbus clouds. A few lightning flashes were observed around the time this picture was taken, so these were indeed thunderstorms. The remainder of the sky has a few cumulus clouds, but they show little tendency for further growth, except possibly over the distant mountains of southern New Mexico. A few cirrus unrelated to blowoff from the cumulonimbus anvils appear in the distance.

Environment: The cumulonimbus clouds reached their peak activity in the afternoon at the time of maximum heating over the desert. This photograph was taken in January when cold temperatures aloft combined with enough surface heating to produce vertical instability resulting in small thunderstorms.

Significance: In this situation, the cumulonimbus clouds are unlikely to grow larger or stronger. Nevertheless, afternoon heating is often sufficient to produce small thunderstorms, so limit outdoor activities that could be adversely affected by lightning. Cumulonimbus clouds like the ones here will probably die before sunset.

Comments: A telltale sign of instability in this dry and cool environment is the cumulonimbus anvil, which here is nearly all that is visible. The fibrous nature of the sky indicates that shower activity will probably stop soon.

SHOWERS AND THUNDERSTORMS

Cumulonimbus clouds invariably lead
to inclement weather. While showers
fall from the cloud base, the positively
charged upper portion of the cloud and
negatively charged lower portion create
electrical discharges that form lightning
and thunder. Although thunderstorms
can cause death and destruction, their
effects may also be beneficial. A
significant amount of the growing-
season rainfall for the central United
States is produced by large showers and
thunderstorms. Lightning oxidizes
atmospheric nitrogen, changing this
plant nutrient into a form more readily
absorbed by the soil.
Rapidly moving narrow lines of showers
or thunderstorms are called squall lines.
Though brief, they are intense and can
be 20–1,000 miles (32–1,600 km) in
length. Similarly short-lived but potent
are microbursts, which look like rising
clouds of dust or descending streams of
rain. Microburst winds can reach speeds
of up to 150 miles per hour (240 kph)
and are particularly dangerous to
aircraft during takeoff.

167–185 Showers and Thunderstorms

Description: *Thunderstorms are local storms produced
by cumulonimbus clouds accompanied by
thunder and lightning.* These storms are
usually accompanied by strong and
gusty winds, rain, and, sometimes, hail.
A shower is precipitation that starts and
stops quickly, changes rapidly in
strength, and results in a large change
in the appearance of the sky—usually
within 15 minutes. In dry regions,
rainfall from a shower may evaporate
before it reaches the ground; such
rainfall is called virga.
A thunderstorm always has a

cumulonimbus cloud within its structure. In drier regions, precipitation from a thunderstorm may not reach the ground, so only virga is seen.

Since showers and thunderstorms grow, mature, and dissipate rather quickly, wind flow, both horizontal and vertical, is much stronger and more turbulent than in large-scale traveling weather systems. Different kinds of severe weather tend to occur during the different life-cycle stages. During the growth stage, strong vertical winds may cause tornadoes, funnel clouds, mesocyclones, wall clouds, and waterspouts as a thunderstorm explodes upward. During the early portion of the mature stage, there may be strong downdrafts in the form of gust fronts, outflows, and microbursts. During the dissipation stage, the cumulative effect of continuing rainfall may cause flash flooding. Lightning can occur during all of these stages.

Environment: Showers and thunderstorms come from cumulus convection, which is due to vertical instability in the atmosphere that is strong enough to allow clouds to grow upward more easily than horizontally. When a cloud grows vertically through thousands of feet of the atmosphere, water often turns to ice and snow in updrafts. Ice and snow often melt into water in downdrafts. Water vapor always forms and evaporates, thereby releasing or taking heat from the air. Hail may form and melt. The horizontal and vertical winds can be strong—sometimes up to 100 miles per hour in any direction.

The reasons showers and thunderstorms form are varied, and they occur in many combinations. Thunderstorms and showers grow upward because of instability in the atmosphere due to the way that temperature and humidity are arranged at different heights. That is, once a cloud begins to go up, it increases its vertical speed until a layer is reached where the temperature-humidity structure stops it from rising. The factor that forces the cloud to grow upward is called the *forcing*. Forcing that is strong enough to result in showers and thunderstorms may be due to a variety of conditions, including the presence of cold- or warm-front boundaries near the ground, mountains and large hills (hundreds of feet or more of vertical rise), or sea or lake breezes along the shores of large bodies of water. Outflow and temperature-humidity boundaries from previous thunderstorms on the same day, or from one day earlier, may also cause forcing. When much colder air than normal overlies warm air, showers and thunderstorms may occur. Internal circulation within thunderstorms may cause further vertical motion that maintains or increases a cloud's intensity. In summer, elevated land masses over which the sun warms the air relative to the surrounding air can cause forcing that leads to a shower or thunderstorm.

Once a shower or thunderstorm has begun as a result of one kind of forcing, it may feed on other kinds of forcing. For example, a cumulus may form almost every day during the summer over a large mountain in the western states or a region along the Gulf of Mexico coast. On some days, the cloud may struggle all day but never reach the shower stage. On other days, the cumulus may quickly feed on strong vertical instability, cold air aloft, or moisture from a traveling weather system that is passing over the region, and develop into a large thunderstorm.

Season: Most thunderstorms occur during the warm months, when the air is moist. They are possible year-round along the Florida and Gulf coasts, whenever enough moisture or warm air is in the region.

Thunderstorm occurrences peak in summer across all of North America except for the Pacific coast, where winter is the peak season. Thunderstorms are normally observed every month of the year over the southeastern quarter of the United States; across most of Canada and the northern plains and mountain regions of the United States, wintertime thunderstorms are virtually unknown.

Range: The geography of the United States is particularly favorable for strong thunderstorms. More than half the world's tornadoes are thought to grow out of severe thunderstorms within the lower 48 states; most occur east of the Rocky Mountains (where several of the forcing factors combine frequently to produce the environment in which strong thunderstorms could occur). In the southern Plains States during the spring months, for example, there may be a cold front, instability aloft, a dry line, an outflow boundary from the previous day's storms, and moist air flowing in from the Gulf—the mixture of conditions may vary widely. Late in the day, some of the storms over the Plains States may grow further into very large convective weather complexes that cover an entire state, with moderate to heavy rain that lasts all night.

Over the mountains of the western United States and the western provinces of Canada, showers and thunderstorms occur nearly every day during the summer because the elevated land is much warmer than the air at the same levels over the plains or valleys nearby, and because of the updrafts produced by the terrain itself.

Along the Gulf of Mexico and Atlantic coasts, showers develop on almost every summer day. Over the Florida peninsula in particular, there is a shower or thunderstorm along the coast every day in summer. In these coastal regions, some showers grow into thunderstorms and move inland for a short or long distance, depending on the availability of moisture, the interaction with traveling weather systems, the intensity and location of the previous day's thunderstorms, and the direction of the prevailing winds.

Variations: The cloud in plate 173 is growing, as shown by its sharply outlined top, which indicates it is still in the updraft stage. Numerous cumulus congestus clouds in 184 are growing, while different wind speeds at varying heights of the atmosphere (shear) are tilting them (the cloud on the left appears to be deep enough to be producing rain). The cloud in plate 168 is producing lightning and is probably still growing, as shown by the rounded bubble elements at the very top. Clouds in plates 169 and 180 are producing lightning while they begin their mature stage, as indicated by the rain, but they still have some flat cloud bases indicative of updrafts aloft. In plate 170, a very large updraft with a partial anvil on top has occurred where large-scale conditions aloft and at the surface are coinciding in one small region. The dark cloud and light rain in plate 167 are beneath a growing cloud to the east in the late afternoon. During the late growth to early mature stage of a thunderstorm, updrafts may become very well organized, as in plates 179 and 182. Toward the mature stage of a thunderstorm's life cycle, a series of smaller updrafts and a larger anvil appear (171). When thunderstorms become very well organized and stretch across several counties, portions of large outflows, shelf clouds, and gust fronts can be seen (174, 176–178, 183). Less significant outflows and downdrafts accompany rain during the mature and dissipation stages (172, 175, 178, 181, 185).

Significance: Showers and thunderstorms can be considered beneficial or detrimental, depending on the circumstances. They produce a significant amount of the growing-season rainfall over the central United States. However, they are also responsible for much of the severe weather in North America.

Comments: In the United States and Canada, showers and thunderstorms are responsible for many of the memorable weather events each year. These situations may include the comfortable sounds of a summer evening's shower with distant, rolling thunder. Or, they may produce a moment of fear associated with a strong gust front that passes over in a matter of minutes and causes damage nearby. It is not uncommon to stop a ball game or call off a project in the back yard when rain and lightning suddenly become too heavy for safety. Activities such as boating, aviation, agriculture, and construction are vulnerable to all of the dangerous effects of showers and thunderstorms.

A vexing problem is the false sense of security that may develop about thunderstorms. After several years without a serious threat from lightning, high winds, or other significant weather, people may fail to react appropriately to storms, which may appear seemingly without warning and can become very dangerous in a short time.

168–169,   **Lightning**
186–193

Description: *Sudden, brief, and brilliant flashes of light, often accompanied by thunder, produced by electrical discharges within and near cumulonimbus clouds.* The flashes may appear as irregular streaks, often with multiple branches, and extend from the cumulonimbus cloud to the ground, to another cloud, or into the air. The lightning flash may also take place entirely within the cumulonimbus cloud, in which case the flash appears as a general illumination of part or all of the cloud.

Environment: The electrical discharge seen as lightning results from the accumulation of electrical charges in various regions within the cumulonimbus cloud, as well as in the air and on the ground in the vicinity of the cloud. For reasons not well understood, when liquid cloud droplets freeze, the ice crystals take on a positive charge and the remaining droplets become negatively charged. Thus, the upper (freezing-temperature) portions of a cumulonimbus are positively charged, and the lower portions are negatively charged. The negative charge in the lower cloud induces a positive charge on the ground beneath the cloud and in the air surrounding the lower cloud, both of which in turn induce a ring of negative charge on the ground outside the cloud. The positively charged upper cloud induces a negative charge in the surrounding air. The electrical discharge occurs between any two oppositely charged regions (with the exception of the area between the oppositely charged ground regions) when the electrical potential gradient between the regions reaches 15 million volts per mile.

The electrical currents unleashed are concentrated in a path several inches across, heating the air almost instantaneously to 20,000°F (11,100°C) or higher (hotter than the surface of the sun). This glowing channel of hot air is visible as the lightning flash, while thunder is the audible explosion of air along the length of the flash caused by the sudden heating. Thunder is usually audible up to 8 or 10 miles (about 13–16 km) from the lightning flash, although, according to recent reports,

it has been heard as much as 70 miles
(113 km) away.

Season: All year in warm climates. Only in
summer where winters are cold.

Range: Lightning occurs at least several times
per year everywhere in North America
except for the Arctic regions of
northern Alaska and Canada. The
frequencies of thunderstorms (any
lightning-bearing storm close enough
to the observer for thunder to be heard)
range from once every 5 or 10 years
along the Arctic coast, to 3 to 10 per
year along the Pacific Coast and
southern Alaska, to 20 to 40 per year
along the U.S.–Canada border, to 60
or more per year across much of the
southern half of the United States. Peak
thunderstorm frequencies in the United
States are 110 per year near Cimarron,
New Mexico, and 130 per year over the
southwestern Florida peninsula.
Lightning-detection systems covering
the United States pick up 40 million
cloud-to-ground strikes per year, or
an average of about 13 per square mile.
In southern Florida, the annual rate
exceeds 30 per square mile.

Variations: The main varieties of lightning are
defined by the locations of the
electrically charged regions involved in
the discharge. Cloud-to-ground
lightning (190, 191, 193) connects the
negative lower cloud with the ground.
The lightning flash begins as an
invisible "leader stroke," then
progresses in steps, roughly 100 feet
(30.5 m) at a time, from the cloud base
to the ground. Often, the leader stroke
branches as it approaches the ground.
The first branch to reach the ground
completes an electrical circuit between
the cloud and ground, and the ensuing
brilliant "return stroke" follows the
jagged, irregular leader path back up to
the cloud base (169, 190). Smaller
return strokes may follow other
branches of the leader (168, 190, 193).
Depending on the amount of charge

available, there may be one or more additional leader and return stroke cycles at intervals of approximately 0.05 second, giving a flickering appearance to the lightning flash. Lightning connecting differently charged regions within a cloud is known as in-cloud lightning, and is several times as frequent as cloud-to-ground lightning. In-cloud lightning may appear as nearly horizontal flashes along the cloud base (191, 192) or may remain within the cloud, causing a general illumination of the cloud (186). Flashes between differently charged parts of different clouds are called cloud-to-cloud lightning. Cloud-to-air lightning (188) reaches from the cloud into thin air. Diffuse flashes in the sky, popularly known as heat (or sheet) lightning, are simply the reflection by clouds and air of ordinary lightning too distant to be seen directly.

Lightning extending from the upper parts of the cumulonimbus to the negatively charged ring of ground surrounding the cloud discharges the positive region of the cloud, and is called a positive flash (187, 189). Positive flashes may have path lengths that exceed 10 miles (16 km), several times the length of ordinary cloud-to-ground flashes, and are proportionately more energetic. Only one ground strike in 30 is positive. Positive flashes are often among the last ground strikes from a dissipating storm, and they are more common in wintertime thunderstorms, perhaps because the positively charged, icy cloud tops are closer to the ground.

Significance: Distant lightning indicates instability sufficient for a thunderstorm to develop, possibly leading to nearer storm activity. Lightning that is close enough for thunder to be heard indicates that a storm may be imminent. The distance to a storm miles) may be estimated by coun the time lag (in seconds) between the lightning flash and the audible thunder and dividing that number by five. The heat of a direct lightning strike may ignite its target, be it a house, oil tank, or tree. About half of all North American forest fires are started by lightning.

Lightning's enormous and rapidly changing electric currents can damage or destroy electrical systems ranging from electronic equipment to human nervous systems. Of the 400 Americans and Canadians struck by lightning each year, about 100 die. On the other hand, there has been speculation that lightning in the earth's primordial atmosphere was one of the factors leading to the origin of life. And although lightning destroys trees by the millions, it also oxidizes atmospheric nitrogen, putting the nitrogen (an essential plant nutrient) into a form that is much more readily absorbed into the soil.

198–201 Squall Lines

Description: *A continuous line of thunderstorms or showers, ranging from 20 miles (32 km) to more than 1,000 miles (1,600 km) in length and usually moving fairly rapidly.* A squall line is often *located along or ahead of a cold front.* The lower leading edge of the storm may appear smooth and laminated (**200**), producing a formation called a shelf cloud. As the squall passes overhead, the dark and turbulent cloud base itself becomes visible (**198, 201**); appearance of the cloud base is followed by a sudden onset of wind and precipitation and a rapid but transient rise in barometric pressure. Wind and precipitation may be intense but are usually brief, and the skies may clear before passage of the trailing cold front.

Environment: Maintenance of a squall line requires a continuous flow of moist air into the squall along its entire length, so most squall lines (even outside the tropics) are found in tropical air masses. The largest and longest-lived squall lines form in tropical air, either along a cold front or 100–300 miles (160–480 km) in advance of it. In the absence of a cold front, gust fronts from several isolated thunderstorms may merge to initiate a squall line. Other squall lines develop in the spiral cloud bands of hurricanes.

Season: East of the Rocky Mountains, the greatest incidence of squall lines is observed in June over the Great Plains from Texas to Minnesota, an area where squalls are extremely rare during winter. In the southeastern states, squall lines may occur during any month of the year. Wintertime thunderstorms are somewhat more likely to be associated with squall lines than are summer thunderstorms, and virtually all squall lines observed in the Pacific states and provinces occur in winter.

Range: East of the Rocky Mountains, the range (and seasonal distribution) of squall lines is similar to that of thunderstorms and lightning; however, despite the high incidence of thunderstorms, squall lines are extremely rare in the mountain areas of western North America.

Variations: Frontal squall lines, those found along or ahead of cold fronts, make up the majority of North American squalls, particularly during the winter and spring. Squall lines initiated by gust fronts are most common where and when fronts are least common: namely, over the southern United States (especially Florida) during the summer.

Significance: Thunderstorms associated with squall lines are likely to be more intense than average, and often bring damaging winds, hail, and intense (but brief) rainfall. Squall lines produce "strai line" winds, which, unlike the rotating winds of tornadoes or the rapidly shifting winds of microbursts, blow from the same direction (most often west or northwest) for the duration of the storm. Small, short-lived tornadoes may occur along the leading edge of a squall line. Owing to the rapid motion and narrow width of most squall lines, rainfall is usually as brief as it is intense, and flooding is relatively rare.

202–205   Microburst

Description: *A sudden, short-lived, localized wind that often appears to radiate outward from a central point.* Microburst winds may become visible as a spreading, then rising, cloud of dust, or as a descending plume of rain that spreads horizontally as it reaches the ground (202–205). During the later stages, the *spreading winds may develop an upward curl* (204, 205). Most microbursts last 1 to 5 minutes and cover an area less than 2.5 miles (4 km) in diameter.

Environment: Microburst winds develop when a strong downdraft of rain-laden air, cooled by evaporation, is forced to spread horizontally as it approaches the ground. Two basic types of microbursts have been identified. *Dry microbursts,* in which *most or all of the rain from a high-based cloud evaporates* during descent through a dry lower atmosphere, are most common in arid climates. In *wet microbursts* (202–205) the *evaporative cooling occurs as dry air is drawn into the rain plume from the side or top of the storm cloud.* Wet microbursts usually are accompanied by heavy rain and are most common in moist climates. Both kinds of microbursts require the presence of dry air in the neighborhood of the rain shower. The rain responsible for microbursts usually originates in cumulus congestus or cumulonimbus clouds; in some dry microbursts the parent cumuliform cloud may be quite small.

Season: Data on microbursts are incomplete and scattered, but suggest that microbursts are most likely during summer afternoons.

Range: It is possible, but not certain, that microbursts are most frequent in areas with high thunderstorm frequencies. Conditions for dry microbursts are most favorable when moist, shower-producing air overruns a drier layer of air, a fairly common situation in the summer, when air flowing inland from the Pacific Ocean is blocked at lower levels by mountain ranges, allowing moist air at higher levels (above 10,000'/3,048 km) to overrun dry air in the valleys of the Great Basin and over the high plains.

Variations: Microbursts may be either wet or dry, as discussed above. The motion of the parent cumulonimbus cloud also affects the nature of microburst winds. Stationary clouds produce microburst winds that radiate outward in all directions from a central point; traveling storms may create microbursts that spread out in a fan-shaped pattern. These fan-shaped patterns may appear as localized intensifications of the straight-line winds seen with squall lines. Rotating thunderstorms, or mesocyclones, may create curving microburst winds, often in the proximity of tornadoes.

Significance: Microburst winds have been measured at as high as 150 mph (240 kph), and their sudden and rapid changes can wreak havoc with vehicles, boats, and aircraft. Airplanes are particularly susceptible during takeoff and landing, when sudden changes in airspeed (speed of the airplane relative to the moving air) may reduce the lifting effect of the wings. Several major aircraft disaster have been directly attributed to microburst winds.

Comments: Microburst winds, even if rotating, are distinguished from tornado winds in that they descend and spread outward, whereas tornado winds converge and rise. However, tornadoes and microbursts may occur in the same thunderstorm, sometimes close together and simultaneously. Strong downdraft-induced winds affecting an area larger than 2.5 miles (4 km) across are known as "downbursts," of which gust fronts are a subtype.

TORNADOES AND OTHER WHIRLS

The most violent type of weather disturbance is a tornado, whose rotating spiral of air sometimes destroys whatever is in its path. Tornado winds may reach up to 250 miles per hour (400 kph), and may pass swiftly through an area or linger in one small location. Dust devils and other smaller whirls are also included in this section, as are waterspouts, which are tornadoes that occur over water.

206, 215–222   Mesocyclone

Description: *A mesocyclone is a cyclone intermediate in size between the large middle-latitude cyclones and the small whirls* such as tornadoes and dust devils. The word *mesocyclone* may refer to *any rotating air circulation between 2.5 and 250 miles (4–402 km) in diameter;* the term most often refers, however, to a rotating thunderstorm (also called a "supercell" thunderstorm) or, more specifically, the intense rotating updraft portion of such a thunderstorm. In a well-developed mesocyclone, the airflow pattern is a smaller version of the circulation observed in a large middle-latitude cyclone, with a low-pressure center, warm and cold fronts, and moist and dry air masses. The rotating updraft of moist tropical air is centered on the lowest barometric pressure, and the warm and cold fronts represent boundaries (mostly gust fronts) between the tropical air and the cooler and drier air descending around the periphery of the thunderstorm. From a distance, a mesocyclone looks like a mature cumulonimbus, having rapidly growing cumulus towers and a well-developed anvil and, sometimes, a noticeable tilt and perceptible rotation.

Because of the peripheral downdrafts, mesocyclones are usually surrounded by clear skies and thus appear in isolation. The characteristic visual features of a mesocyclone are the *flanking line of cumulus congestus* clouds that usually extends *southwest from the main cumulonimbus;* the *extensive, flat, rain-free base* (indicating strong updrafts) (216, 220) beneath the flanking line and main cloud; and the *wall cloud* (206, 215, 218–222)—a distinctive, localized, and frequently rotating lowering of the rain-free base. In mesocyclones the precipitation area is usually north or northeast of the rain-free base (215, 217, 218, 221), often immediately adjacent to the wall cloud.

Environment: Mesocyclones are relatively rare and require a very special set of meteorological conditions. Among these are a moist, low-level air mass and drier air aloft; strong instability; a strong jet stream aloft; a change in wind direction and speed between the lower and upper levels of the troposphere; and some preexisting rotation in the lowest layers of air. All of these conditions may be found in the warm sectors of strong cyclones, and a sufficient number of these conditions to generate supercells can sometimes be found elsewhere.

Season: Chiefly spring and early summer, but may occur at any time of year if conditions are correct.

Range: Most often reported from the Midwest and Plains States; may occur anywhere that ordinary thunderstorms occur.

Significance: Mesocyclones frequently produce severe weather, including strong winds, large hailstones, and tornadoes. Tornadoes occurring beneath the flanking line of cumulus clouds are generally small and weak; the rotating updraft associated with the wall cloud sometimes becomes a large tornado. When the conditions for mesocyclone formation are observed, the National Weather Service may issue a "Tornado Watch," indicating that tornadoes may form over a restricted area during a given time period. A "Tornado Warning" indicates that a tornado has formed or is imminent. The development of a rotating wall cloud beneath a rain-free base is the most reliable visual indicator that a tornado is likely to strike a given location within minutes. The air circulation of a developing mesocyclone may also be detected by Doppler radar (which can measure the speed of air moving toward or away from the radar, in addition to the intensity of precipitation), providing as much as 20 minutes' advance warning before a tornado touches down.

207, 209–214, 223–230	Tornado
Description:	*A rapidly rotating column of air extending from the base of a cumulus congestus or cumulonimbus cloud to the ground.* The column may be made visible by the presence of dust and debris, or by the condensation of water vapor due to the lowered pressure within the rotating column. Tornadoes may range from several feet to 2 miles (3.2 km) in diameter, may last from several seconds to as long as 7 hours, and may track along paths from several yards to more than 200 miles (320 km) long.
Environment:	The largest and most dangerous tornadoes occur in association with mesocyclones (supercell thunderstorms). They require the same meteorological conditions as do mesocyclones: a moist, low-level air mass and drier air aloft; strong instability; a strong jet stream aloft; a change in wind direction and speed between the lower and upper levels of the troposphere; and some preexisting rotation in the lower layers of air. A powerful spring cyclone

creates widespread conditions favorable to the formation of tornadoes; such circumstances may lead to a tornado outbreak, in which numerous mesocyclones spawn tornadoes over a large area. Smaller and weaker tornadoes are often observed in association with squall lines and gust fronts, beneath rapidly growing cumulus congestus clouds, and in hurricanes. Development of these weaker tornadoes requires strong instability, leading to strong updrafts, and some preexisting rotation in the near-surface winds. By drawing the rotating surface air upward, the updraft intensifies the spin (a process resembling an inverted version of water going down a drain).

Season: Tornadoes have occurred in every month of the year; over all, April and May are the peak months. Early in the year the center of tornado activity is along the Gulf Coast, but as spring progresses, the action moves north. The tornado belt reaches the northern Plains States, southern prairie provinces, and New England by July, and after August shifts back south.

Range: The Great Plains provide the most favorable environment on earth for the formation of tornadoes. Seventy-five percent of the world's reported tornadoes (about 1,000 annually) occur in the United States, and another 5 percent touch down in Canada. Tornadoes have been sighted in every state and province of the United States and Canada, including Alaska and the Yukon. Central Oklahoma has more tornadoes per acre than any other place on earth.

Variations: Tornadoes come in a variety of sizes and shapes; they usually change dramatically in appearance during their lives. The most powerful tornadoes, with winds possibly exceeding 250 mph (400 kph), emerge from mesocyclones. These supercell tornadoes can endure for an hour or more, and a persistent mesocyclone may spawn several. The first indication of an incipient tornado is often a dimple-shaped lowering from a wall cloud or rain-free base (220). As the rotation intensifies, the dimple extends earthward, evolving into a conical funnel cloud (225, 226). The funnel becomes a tornado when it reaches the ground (227) or when a rotating cloud of dust and debris on the ground beneath the funnel (209) rises to meet the descending funnel, indicating that the rotating air column has reached the ground (210). After reaching maximum strength (211, 212), a tornado narrows and the funnel becomes distorted (213), sometimes developing a ropelike shape immediately prior to its dissipation. Ropelike funnel clouds may undergo fantastic contortions (214), and the lower end of the funnel may even return to the cloud base (207). The powerful updrafts inside a tornado can suspend tremendous loads of dirt and debris. When the tornado weakens or dissipates, the debris cloud may suddenly collapse earthward, spreading horizontally away from the dissipating tornado (214) much like a microburst.

Many non-supercell tornadoes form when updrafts in a rapidly developing cumulus congestus or cumulonimbus cloud draw in slowly rotating low-level air. This mechanism is similar to that which forms many waterspouts, leading to the colloquial name "land spout" for a non-supercell tornado. The preexisting slow rotation of the low-level air can be caused by air flow around mountains or ridges or the convergence of sea breezes or gust fronts. The rotation of many non-supercell tornadoes begins near the ground and grows upward (the opposite of the mechanism in supercell tornadoes). Consequently, the first evidence of a non-supercell tornado is often a dust whirl on the ground (229, 230), which forms a funnel cloud only after the rotation has reached cloud base (228). Non-supercell tornadoes are typically small, short-lived (5–15 minutes), and weak, with winds that rarely exceed 120 mph (193 kph); as a rule they do far less damage than supercell tornadoes. Most tornadoes over the Florida peninsula and over the High Plains within 100 miles of the Rocky Mountains—places with large numbers of mostly weak tornadoes—are the non-supercell variety.

Tornadoes in hurricanes are usually small and short-lived. Most occur soon after a hurricane makes landfall, in the heavy rainsqualls where the storm winds blow inland (to the right of the eye, as you look in the hurricane's direction of motion). In 1967, Hurricane Beulah unleashed 115 tornadoes on southern Texas, most of which were located north of the hurricane's westward path. It is likely that many tornadoes in hurricanes go unreported because they are masked by heavy rain and their damage is obscured by the general destruction of the storm.

Strictly defined, a tornado is a rotating air column extending from ground level to cloud base. Thus, a funnel cloud extending only part of the distance from the cloud to the ground, with no evidence of rotation on the ground, is not a tornado (223, 224). Likewise, a rotating dust cloud on the ground with no evidence of rotation at cloud base is, strictly speaking, not a tornado. However, in some cases a complete (although weak) rotating column may exist but not be visible, because of a lack of ground debris or insufficient atmospheric moisture. Small whirls that develop along the leading edge of a gust front are frequently seen as rotating, vertical plumes of dust (229).

Although these have been called gust front tornadoes, or gustnadoes, it is possible that relatively few have rotation extending to cloud base. There are many kinds of whirls, ranging from supercell tornadoes to dust devils and including hybrids (such as gustnadoes) that may or may not be tornadoes. In coming years the definitions of tornadoes will need to be revised to accommodate the expanding variety of tornadolike whirlwinds.

Narrow streamers of rain can, at a distance, look like tornadoes. Turbulent and threatening scud clouds may be ruled out as tornadoes because of their lack of rotation. Horizontal shelf clouds are sometimes erroneously reported as tornadoes.

Significance: Over the past century, tornadoes in North America have killed between 15,000 and 20,000 persons; deaths have decreased in recent years thanks to improved forecasting and warning systems. During the decade of the 1980s, the human loss in the United States and Canada averaged about 60 per year. Meanwhile, property damage due to tornadoes has increased steadily in most years since 1973, to an annual loss of $1 billion or more.

Comments: Most tornadoes rotate cyclonically: counterclockwise in the Northern Hemisphere and clockwise in the Southern Hemisphere. Most of the few anticyclonic tornadoes recorded were in conjunction with nearby cyclonic tornadoes.

Large whirls often spawn smaller whirls. Thus, mesocyclones are found in large-scale cyclonic storms, and tornadoes are born within mesocyclones and hurricanes. Small but intense whirlwinds have also been observed within tornadoes, traveling in a circular path around the main whirl. These tiny suction spots (named for their action on objects on the ground) are typically less than $30'$ (9.1 m) across and last only seconds, but they are responsible for some of the remarkably erratic patterns of destruction observed in "multiple-vortex tornadoes." Suction spots are dramatic as they writhe, snakelike, within the main tornado circulation; in still pictures they appear as streamers and funnel-shaped concentrations in the debris cloud (229).

231, 232 Waterspout

Description: *A waterspout is a tornado over water; a rapidly rotating column of air extending between the base of a cumulus congestus or cumulonimbus cloud and the surface of a lake, river, or ocean.* Lowered pressure within the rotating column condenses water vapor in the air and lifts it from the water surface, making the column visible (231, 232). Waterspouts are *not* filled with liquid water. The cloud of spray around the base of a waterspout, analogous to a tornado's debris cloud, is called the bush. Because of the ready supply of water vapor, complete cloud-to-surface funnel clouds are more likely with waterspouts, but sometimes the bush and a small funnel cloud aloft provide the only evidence of the rotating air column (232). Some waterspouts have a broad, nearly conical collar cloud, where the tubular funnel cloud enters the main cloud base (232); collar clouds are also seen with some tornadoes. Some tornadoes become waterspouts when they move offshore or across rivers or lakes; waterspouts may turn into tornadoes when they cross land. However, most waterspouts remain entirely over water and, in general, are smaller and weaker than tornadoes.

Environment: Supercell, or tornadic, waterspouts form under the same conditions that bring about supercell tornadoes: a moist, low-level air mass and drier air aloft; strong instability; a strong jet stream aloft; a change in wind direction and speed between the lower and upper levels of the troposphere; and some preexisting rotation in the lowest layers of air. Waterspouts are most common in unstable, moisture-laden air overlying warm tropical and subtropical ocean surfaces, particularly where the water temperature is 80°F (27°C) or higher. The large majority of obscured waterspouts, the non-supercell waterspouts, develop when slowly rotating air near the surface is drawn up by the strong updrafts in a rapidly growing cumulus congestus or cumulonimbus cloud.

Season: Observed chiefly during summer months.

Range: Because waterspouts are most likely to occur over water 80°F (27°C) or warmer, the Gulf of Mexico and the Atlantic Coast from Florida to Cape Hatteras, North Carolina, are prime locales. In North America the number of observed waterspouts is greatest in the vicinity of the Florida Keys during summer. Waterspouts probably occur over all tropical water, but most sightings are of those occurring within 10 miles of land; converging air currents and weak rotating eddies in the lee of small islands and around irregularities in the coast may actually increase the number of waterspouts in near-shore areas.

Variations: There are two main types of waterspouts: supercell (or tornadic) and non-supercell. Non-supercell waterspouts that form beneath unimpressive cumulus congestus clouds in partly cloudy skies are called fair-weather waterspouts. Supercell waterspouts are in every way identical to supercell tornadoes, except that they form over water, a fact that may make it easier for them than for their land counterparts to start spinning.

Significance: Waterspouts present a threat to shipping and coastal facilities. However, because of the maneuverability of boats and ships and because fair-weather waterspouts usually dissipate quite rapidly when they come ashore, the threat of damage is much less than that posed by tornadoes on land.

208, 233–236   Lesser Whirls
Dust devils, snow devils, steam devils, etc.

Description: A wide variety of *small, mostly short-lived whirls seen over* surfaces such as *dusty ground, water, and snow*; sometimes seen in midair.

Environment: Lesser whirls of all descriptions originate in a preexisting rotation in the atmosphere that is caught in an updraft. Usually, but not always, the updraft results from strong, local instability in the lower atmosphere. The combination of rotation and updraft leads to a process called *vortex stretching:* The vortex (funnel of rotating air) is stretched vertically by the updraft, forcing the vortex to become narrower and, therefore, spin faster. The sources of the rotation and of the instability vary, although frequently the initial rotation can be attributed to nothing more than random turbulent eddies in the air. Dust devils may derive their initial rotation from winds blowing around subtle terrain features. The instability of the air comes from intense heating of the ground by the sun during late morning and early afternoon.

Season: Dust devils are particularly common in late spring and summer. Steam devils appear in winter over lakes, or at any time of year over warm water with overlying cold air. Snow devils occur in winter. Wing-tip vortices and other small whirls may occur at any time of year if conditions are right.

Range: Dust devils are seen most commonly over deserts and dry scrubland, less frequently elsewhere. Steam devils are seen throughout the continent. Snow devils occur most often in high mountain areas. Wing-tip vortices and other small whirls can occur anywhere if conditions are right.

Variations: The most famous of the lesser whirls is the dust devil, a dust-filled whirlwind that can extend up to tens of yards across and hundreds of feet high. The largest recorded dust devils (spotted in Utah) were 2,000' (610 m) high, lasted 7 hours, and wandered 40 miles (64 km) across the desert. The typical dust devil occurs under a cloudless or partly cloudy sky (208, 234). Cumulus congestus clouds are sometimes seen above dust devils (233); on rare occasions the dust devil's rotating updraft merges with that of a growing cumulus congestus cloud to create a hybrid non-supercell tornado.

Steam devils derive their instability from the temperature contrast between cold air overlying warm water—a condition existing, for instance, at a hot spring in Yellowstone National Park (235, 236) or at heated ponds near power plants. The vortex receives its spin when steam is ejected from a geyser or comes up off the warm pond. In some steam devils, both ends of the rotating column are in midair, and it appears that the two ends may connect to form a doughnut-shaped rotating ring. Somewhat larger steam devils develop over lakes in the winter; during arctic air outbreaks over the Great Lakes, when frigid air overruns the relatively warm lakes, steam devils have reached heights of several hundred feet. Snow devils are fairly frequent in high-mountain country on winter days. The initial rotation of a snow devil begins as the wind blows around ridges. The vortex is stretched vertically as it moves from the ridgeline over downward-sloping terrain.

Wing-tip vortices are spun off from both sides of airplanes; they sometimes appear as rolling clouds of dust during landing or takeoff and in the sky as parallel tubular structures in contrails (visible with binoculars). The rotation results from the shape of the wings. Air blowing across the curved upper wing surface generates a low-pressure zone (or partial vacuum) that creates lift; the difference in pressure between the top and bottom of the wing forces air to curl upward around the wing tip, creating the rotation. The heavier the airplane, the greater the pressure difference and the stronger the generated vortices. With large aircraft, such as commercial jetliners, the vortices are so intense that they remain visible in the contrail for several minutes.

Significance: The vast majority of the lesser whirls are completely harmless—one can run through most dust devils without suffering more than a faceful of dust. A few extraordinary dust devils, however, have winds reaching 90 mph (153 kph), equivalent to the velocity of a weak tornado, leading scientists to conclude that some may, in fact, *be* weak tornadoes. Strong dust devils have overturned parked airplanes and taken the roofs from buildings. Dust devils beneath growing cumuliform clouds are the most likely to become dangerous. The main effect of wintertime lake steam devils is transport of moisture from the lake surface into the cold air mass, contributing to lake-effect snowfalls on the downwind shore. Snow devils can be annoying to skiers and can flatten the tents of winter campers. Wing-tip vortices from large aircraft are capable of overturning smaller airplanes, and are calculated into aviation experts' flight regulations.

HURRICANES AND TROPICAL STORMS

Born as tropical storms, hurricanes are cyclones with a calm central eye and a surrounding wall of wind, rain, and clouds. When fully developed, they can exceed wind speeds of 155 miles per hour (250 kph). Hurricanes require an entire day or more to pass over a given area; they sometimes cause great damage.

237–241 Hurricanes and Tropical Storms

Description: *Hurricanes (winds of 74 mph/119 kph or more) and tropical storms (winds of 39–73 mph/63–117 kph)* have distinctive features that separate them from most other types of storms. *Large storms in the tropics covering a large area* and *lasting several days* without major changes in their structure are broadly called *tropical cyclones,* a term that also includes tropical depressions (winds under 39 mph/63 kph). All hurricanes proceed through the sequence of tropical depression and tropical storm before achieving designation as a hurricane. Often during a hurricane or tropical storm there are mixed layers of cumulus and cumulonimbus clouds in the cloud shield of the major storm. From the ground they sometimes appear as long, curved lines or rows of clouds—the rain bands—at various atmospheric levels. These clouds stretch across the sky. All hurricanes have an "eye"—an area of nearly calm winds in the middle of the eye wall, where the storm's strongest winds blow in a circular band that borders the eye. It is near the eye that very heavy rainfall often occurs, many times exceeding 10″ (25 cm) or more.

As a result of the storm's strong effect on the surface of the ocean, its bays, inlets, canals, bridges, islands, and shorelines are vulnerable to catastrophic flooding due to the storm surge and wave action.

Because of their size, hurricanes and tropical storms take a day or more to completely pass over a given place.

Environment: For a tropical cyclone to form, the temperature of the ocean should be above about 78°F (25.5°C). At these temperatures, adequate heat from condensing water vapor produces cloud growth and leads to the unstable conditions that permit the storm to become more intense. Also, at these temperatures the moisture content of the air near the ocean's surface is very high.

A tropical cyclone can form only in areas more than about 10° latitude from the equator, where the earth's rotation is strong enough to translate into a rotating storm. The causes may be a preexisting traveling disturbance, an upper-level trough, or sometimes a very complex combination of wind and temperature patterns at all levels. These specific conditions for storm formation can often be identified and forecast successfully by highly specialized meteorologists using the most recent data from the tropics.

Season: May–November for tropical cyclones that form in the Atlantic, Gulf of Mexico, and Gulf of California, and that may later affect the United States and Canada.

Range: The strongest tropical storms and hurricanes are most likely to occur in the late summer along the Atlantic and Gulf of Mexico coasts. Within the first hundred miles of the coast where the tropical cyclone comes ashore, the maximum winds, heaviest rainfall, and highest storm surge are felt near the eye.

Over the North American land mass, the impact of tropical cyclones can extend inland for a great distance. In the Southern, Eastern, Atlantic, and New England states of the United States, and the Maritime Provinces of Canada, the heavy rain (more than 5″/13 cm) and strong winds (more than 50 mph/80 kph) can extend for hundreds of miles inland from the landfall position. The resulting rainfall is often considered beneficial if it is not extreme in amount, duration, or intensity. Over Texas and the southern Plains States, rainfall and clouds from storms entering the Texas coast may bring substantial to excessive rainfall. Southern California and Arizona, and to a lesser extent the states to their northeast, receive some rain and clouds every year from tropical storms or hurricanes, or their remnants, during the late summer to fall months.

Variations: A dramatic example of the combination of winds and water resulting from a tropical cyclone as it comes ashore is shown in plate 237. Notice in plate 237 the gray nimbostratus sky and the low visibility due to heavy rain. Plates 238 and 240 have the same characteristics in somewhat weaker conditions. During a hurricane there are often lulls in the wind and rain between the very intense rain bands or in the eye itself. Since visibility is often poor, these bands are hard to see as they approach. Wind and wave conditions can worsen in a matter of minutes, with few visible warning signs in the sky. A turbulent sky (239) may be seen just before a rain band of a tropical cyclone passes overhead. Such a sky can be accompanied by strong and very gusty winds over wide areas, making aviation and boating activities virtually impossible. In addition, tornadoes form in some hurricanes and tropical storms, after they have made landfall, from skies like the view in plate 239, rather than from skies in the classic views from individual thunderstorms. The view in plate 241 clearly shows the linear structure of the outer reaches of a tropical cyclone. It also shows the multiple layers of clouds that are often stacked above each other, producing the very dark sky along the horizon. Imbedded cumulus are also apparent in plate 241.

Significance: The damage from tropical cyclones ranges from complete devastation, caused by the passage of the eyewall of a very intense hurricane along the coast, to a minor nuisance, produced by a weak tropical storm whose effects resemble those of a strong thunderstorm. In assessing the threat from tropical cyclones, both water and wind impacts are important.

The greatest damage from tropical cyclones is from water effects. The storm surge is often close to 10′ (3 m) high or more for a hurricane, and is particularly important because it causes a general rise in the level of the ocean as the storm approaches a bay, inlet, island, bridge, or other coastal feature or structure. On top of the storm surge is the daily tide, which may increase the storm surge by several more feet. Finally, over this general rise are waves ranging from 25′ (8 m) or higher (237) during the most intense storms.

Damage from winds increases as the square of the speed; that is, the speed multiplied by itself. For example, the force of the wind in a weak hurricane (74 mph/119 kph) is doubled when the wind increases to about 100 mph (an increase of the wind's speed by only one-third) . Such information is important in order to take precautions against the dangerous effects of high wind. Precautions against falling trees and other windblown objects, power lines, and other structures need to be anticipated when the forecast projects such strong winds. In many locations in the United States, winds over 50 mph (80 kph) occur almost every year, and not too infrequently they reach 75 mph

(121 kph) without causing much damage. However, since a 100-mile-per-hour wind doubles the damage, it is potentially much more significant.

Comments: Most of the United States is affected in some way every year by the direct or indirect wind and water effects of tropical storms and hurricanes. The U.S. coastline is becoming increasingly populated in urban areas, and a large proportion of these people are not familiar with the strength and effects of tropical cyclones. While the most common experience is of the weaker tropical storm and its relatively minor flooding and winds, the potential exists for extreme damage along the coasts, and for severe flooding inland under specific conditions.

242–255 Satellite Views of Hurricanes and Tropical Storms

Description: The fully formed hurricane photographed from a satellite usually shows several distinctive features: the eye in the center, where the wind is relatively calm (under 25 mph/40 kph); the large cirrus-cloud shield, a circular, spiral shape swirling outward from the eye; and, on the outer edges of the storm, if not hidden by the cloud shield, rain bands spiraling into the center. In tropical storms, the eye is often missing, the circular cloud shield around the center is usually not complete, and there are fewer visible rain bands than in a hurricane.

Environment: The circulation patterns associated with tropical cyclones are often shown clearly in satellite views of hurricanes. In the lower levels near the ocean surface are the rain bands spiraling inward toward the eye. At the top of the storm is the cirrus-cloud shield's overcast region, showing an outward spiral. In a well-developed storm, these clouds hide many of the rain bands below. These high clouds represent the outward flow of air aloft that has been brought inward at low levels, has moved upward in the eyewall, and now must move outward to continue the circulation. Any breakdown in these basic features will cause a weakening of the entire storm system.

Variations: These spectacular views of tropical cyclones were photographed by astronauts using hand-held cameras. The entire structure of spiral bands flowing inward, the eyewall, and the cirrus clouds flowing outward are shown in plates 242 (Typhoon Pat, taken from the satellite *Discovery*) and 244 (Typhoon Gladys, taken from *Apollo* 7). Detailed views of the center of a hurricane are shown in plates 243 (Hurricane Elena, taken from *Discovery*) and 245 (Typhoon Pat, taken from *Discovery*). Plate 245 also shows the eyewall clouds penetrating above the general cirrus-cloud shield, and the turbulent circulation in cumulus clouds in the eye near the ocean.

Plates 246–249 show specially enhanced views of Hurricane Hugo, the very intense hurricane that principally hit the Virgin Islands, Puerto Rico, and South Carolina during September 1989. As the storm approached the South Carolina coast, meteorological instruments detected different features of the hurricane. Plate 246 shows Hugo's general features, using enhanced infrared satellite imagery, when the eye was over Charleston. Plate 247 shows Hugo's circulation using a method where the lowest clouds are dark and the highest are white. Plate 248 shows an electronically enhanced and color-coded view of Hurricane Hugo taken by the National Weather Service radar at Charleston. A method that codes cold (high) cloud tops in color, with red as the coldest, was used to enhance the infrared return of the storm in plate 249; it vividly shows the storm's eye.

A sequence of four stages of the development of Hurricane Roslyn, a tropical storm that began in the Pacific Ocean off Mexico in October 1986, is shown in plates 250–253. The first view (250) shows the tropical-storm stage, when the system has some low-level wind flow toward the center of the storm and a nonsymmetrical outflow aloft. In the next view (251), the inflow stage and eye are well developed, but the outflow is weak, allowing the circulation of the rain bands to be clearly visible. In plate 252 the cirrus outflow is well established and hides most of the low-level structure, except the eye, which is still visible. In plate 253 a large cirrus area is apparent to the north and east, but the low-level circulation remains in a small area to the southwest of most of the cloud cover. Such a situation usually indicates that the entire storm circulation is becoming disorganized and decreasing in strength.

Details of the cumulonimbus tops of rain bands penetrating the outflow cirrus shield of Hurricane David are shown with visible-channel imagery in plate 254. The view of the small but intense Hurricane Diana in plate 225 shows the eye and, to the west, low-level circulation over North Carolina.

Comments: Meteorological satellites have provided unprecedented data on the general organization, the detailed structure, and the motion of tropical cyclones. The availability of such data every half hour over the hurricane-prone regions along the U.S. coasts has become a major resource for tracking and warning of tropical cyclones by the National Weather Service, especially through its National Hurricane Center in Coral Gables, Florida.

SNOWSTORMS AND ICE STORMS

One or more ice crystals born of atmospheric moisture make a snowflake; when snowflakes form in significant number and become heavy enough to fall from the clouds, a snowstorm is born. Although snowstorms can cause damage, they also provide valuable precipitation needed to sustain human, animal, and plant life. Ice is also important in maintaining the balance of the ecosystem, acting as an insulator when floating on the surface of water.

256–284   **Snow**

Description: Snow consists of *large* and often *complex crystals that originate in clouds and fall to earth*. It is perhaps the most ubiquitous variety of ice. Snowflakes can be composed of single ice crystals or large, multi-crystal aggregates, which are seen especially in heavy snowfalls. Aggregate snowflakes grow as large as 2″–3″ (5–7.6 cm) across. The structure and size of ice crystals depend on the temperature and moisture content of the air in which the crystals form. Crystals that grow from the meager water supply at −20°F (−29°C) or below form pencil-shaped hexagonal columns. At temperatures of about −10°F (−23°C) to 0°F (−18°C), most crystals are flat hexagonal plates. Warm air contains more moisture than cool air, allowing larger crystals to grow. At temperatures of 0°F–20°F (−18°–7°C), crystals become large, delicate, six-pointed dendrites (from the Greek for "branched"). Temperatures near the ground are typically 15°–20° higher than cloud temperatures, so dendrites are usually seen falling when the surface

531

temperature is about 15°F–20°F (about −9°C–7°C). From 20°F–32°F (−7°C–0°C) crystals grow into splinter-shaped "needles."

Environment: Most snow forms in supercooled water-droplet clouds, such as nimbostratus, cumulus congestus, or cumulonimbus. Under extremely cold conditions small ice crystals can grow directly from sublimation of water vapor, without the presence of any clouds. Snow can often be seen falling from high clouds and evaporating long before reaching the ground. The clouds responsible for snow can result from frontal cyclones, convection, or orographic forcing.

Season: Winter is the snowiest season for most of North America. In some interior and arctic locations, however, such as the eastern Rocky Mountains and northern Alaska, cold, dry air masses prevail during midwinter; there, spring and autumn are the snowiest times of year.

Range: In North America, snow is rare in some places and inevitable in others. The only area of the mainland United States that has never seen snow in recorded history is extreme southern Florida (south of Miami) and the Florida Keys. San Diego, California, and Yuma, Arizona, have never seen measurable snow (0.1″/2.5 mm or more) but have reported flurries on several occasions. Most of the Hawaiian Islands have never recorded snow, except for the highest summits of Maui and the Big Island of Hawaii. Snow is infrequent in Mexico, although the northern deserts of Chihuahua and Sonora may receive light snowfall once or twice a year. Mexico City, at 7,340′ (2,237 m) elevation, has recorded several measurable snows, and the higher volcanoes in central Mexico have permanent snow fields. Most of North America receives from 1″–80″ (2.5–203 cm) of snow per year, a rough average for the continent being about 40″ (102 cm).

Average snowfall generally increases to the north and with higher elevations, but there is a great deal of local variability. The highest annual average snowfalls are recorded in the coastal mountains of Washington, British Columbia, and southeastern Alaska, where the combination of frequent cyclones, persistent moist onshore winds, upslope forcing by the mountains, and cooling due to elevation allows seasonal totals of 500″ (1,270 cm) or more. In the high Arctic, however, annual snowfall averages 20″–40″ (51–102 cm), about the same amount as New York City.

Records: Extreme seasonal snowfall records are 975″ (2,477 cm) at Thompson Pass, Alaska, in 1952–53, and 1,122″ (2,850 cm) at Paradise Ranger Station on Mount Rainier, Washington, during 1971–72. The presence of glaciers at both locations attests to the heavy snowfall.
In eastern North America, the heaviest seasonal totals approach 200″ in eastern Quebec and Newfoundland. The summits of Mount Logan, Quebec, and Mount Washington, New Hampshire, each average more than 250″ (635 cm) of snow per year, and the eastern North America record for one winter is 565″ (1,435 cm) at Mount Washington in the winter of 1968–69. Other snowy areas are east (downwind) of the Great Lakes, where heavy lake-effect squalls can deposit several feet of snow in 24 hours. Average yearly totals range as high as 207″ (526 cm) in the snow belt east of Lake Ontario; average totals of 100″ (254 cm) or more are recorded along the eastern shores of the other Great Lakes.

Variations: Large aggregate snowflakes, usually composed of dendrites, stick to each other and to any horizontal surface; they accumulate even on small surfaces such as tree branches (256, 257). The smaller crystals that fall at surface temperatures below 15°F (−9°C) are much less adhesive and generally do not accumulate very well on small surfaces such as trees (261, 262). At temperatures near or above freezing, partial melting allows snowflakes to adhere to vertical surfaces (259). Winds with speeds greater than 20 mph (32 kph) can return fallen snow to the air, creating drifting and blowing snow. By definition, *drifting snow remains within 5'(1.5 m) of the snow surface* (265); it does not strongly affect visibility at eye level, but it can impede motor traffic by depositing snow on highways faster than plows can remove it. *Blowing snow is raised more than 5' above the ground* (266) and is usually *accompanied by winds in excess of 35 mph* (56 kph). A snowstorm is defined as a *blizzard* when *heavy falling snow combines with blowing snow and winds of 35 mph or higher* to reduce visibilities to near zero. *Heavy blowing snow with strong winds in the absence of falling snow is called a ground blizzard* (266). Scouring and deposition of snow during episodes of blowing and drifting snow can leave drifts and pits in the snow surface (283, 284). These snow formations are known by their Russian name, *sastrugi*. During warm weather, snowdrifts sometimes develop a hard crust (284).

Avalanches (258) are a *common aftermath of excessive snow accumulation on mountain slopes,* a situation often exacerbated by snow blowing over ridges to accumulate in cornices and unstable layers on the lee sides of ridges. Many avalanches are triggered by collapsing cornices or by layers of snow breaking free along their bases. *Wet, cohesive snow layers slide downhill as slab avalanches; looser snow usually disintegrates into cloudlike powder avalanches* (258). The heavy, snow-laden air of a powder avalanche can reach speeds of 200 mph (322 kph).

Significance: Snow is of tremendous importance in air-climate patterns, and it has a ma' effect on our daily lives. Perhaps the best way to appreciate the significance of snow is to consider what happens when it falls in places not accustomed to it, and what happens when it fails to fall in places that rely on it. Snow provides water for forests, agriculture, industry, and drinking. (In California, for instance, most of the water for irrigation comes from snowmelt.) Insulating and reflective snow cover on the ground can affect the global climate and the local weather.

267–282   Ice

Description:
Simply stated, *ice is frozen water*. Ice in the atmosphere and on the ground can assume any of a multitude of forms, depending on the circumstances under which water is converted into its solid state. A water molecule consists of two hydrogen atoms attached to a single, larger oxygen atom; the angle between the hydrogen atoms is $120°$. This is the same angle as the angles of a hexagon, accounting for the characteristic six-sided structure of ice crystals.

Ice can form in the atmosphere and fall to the ground as snow or hail (279); ice can also form directly on the ground. The freezing of liquid water produces ice, but ice can also form by sublimation, that is, directly from individual molecules of water vapor (which is a gas). Ice formed by sublimation displays a distinctive crystalline structure (275, 276); crystals in ice frozen from liquid water are usually so small that the ice appears amorphous (277).

Environment:
Ice requires only two conditions: the presence of water, and temperatures below freezing. However, some varieties of ice, notably frost and black ice on highways, form when the air temperature is slightly above freezing

but the ground temperature is below freezing. Evaporation on wet surfaces that are exposed to above-freezing air temperatures can lower the surface temperature below the freezing point; this effect is seen, for instance, when wet laundry freezes solid at an air temperature of 35°F (17°C). Conversely, water does not always freeze immediately when its temperature falls below 32°F (0°C). Undisturbed pure water can remain liquid at temperatures as low as −40°F (−40°C), and "supercooled" water is observed in clouds. At temperatures between 32°F and −40°F, water must come into contact with solid particles that act as "freezing nuclei" to initiate freezing. On the ground, the inevitable impurities and disturbances assure that water freezes at 32°F (0°C); seawater freezes at 28°F (2.2°C).

Season: Late autumn, winter, and early spring. At higher elevations and in the Arctic, ice also forms during freezing weather in the summer. On very large bodies of water (such as the Great Lakes), ice may not form until midwinter to late winter because several months of winter temperatures are needed to cool the water to the freezing point.

Range: Most of North America; absent from southernmost coastal Florida, Florida Keys, and Channel Islands off southern California.

Variations: On small ponds and puddles, ice first freezes in a thin layer with definite crystal structure (278). As the ice thickens, the crystal structure becomes less apparent (277), in part because of air bubbles trapped in the ice. On lakes large enough to have waves, the first ice to form is a thin surface layer of slush, sometimes called grease ice or frazil ice. Eventually the surface ice grows into small floes of pancake ice (268, 270); if the lake is small enough or the weather persistently cold enough, the floes may freeze together into a fairly solid sheet of pack ice (269). Pack ice may cover the entire lake or be restricted to near-shore areas. During cold and windy weather, spray may coat ships, offshore installations, and shoreline structures with massive loads of ice (267, 269, 272). Along the shoreline, a continuous rain of freezing spray sometimes builds up a wall of ice, grounded to the beach (268). The icicle is a common form of amorphous ice; icicles commonly form from dripping water, but they can also grow from groundwater emerging into subfreezing air (274). Rain falling through a layer of subfreezing air near the ground and onto cold objects may coat everything in sight with glaze or freezing rain (271, 273). Ice storms occur when glaze accumulates an inch or more in thickness and causes damage.

Ice formed by sublimation—that is, directly from water vapor—is hoar frost. Commonly observed varieties include window frost (276) and ground frost (275, 280, 281). The temperature of the object must be lower than the dew point (the temperature at which water vapor condenses), and the air temperature must be higher than the dew point. Thus, the window or ground must be colder than the air (and also be below freezing) for frost to form on it. The window is chilled below the dew point of the indoor air by losing heat to cooler outdoor air. The ground and plants cool by nocturnal radiation (which is why frost occurs on clear nights; a heavy cloud layer at night blocks the escape of radiation, slowing nocturnal cooling). Water vapor condenses and sublimates more readily onto edges and corners than onto flat or concave surfaces, so the fringes of leaves, for example, gather frost more readily than do the centers of the leaves (280, 281). Particularly heavy ground frosts (275) often receive moisture from underlying damp soil. Window frost has slightly curved crystals (276) because organic impurities on the window interfere with crystal growth. Rime (282) consists of supercooled fog droplets that freeze onto exposed objects such as wires and trees. Sometimes local conditions, such as the fog produced by a hot spring, create rime (260).

Comments: Unlike most substances, water expands when it freezes, an odd physical property that has profound consequences. Were ice denser than liquid water, it would sink. Without the insulating effect of floating ice sheets, surface water would lose heat more rapidly; if ice did not float, some large bodies of water, such as the Arctic Ocean, Hudson's Bay, and perhaps some of the Great Lakes, might eventually freeze up completely. Summer thaws would be confined to a thin surface layer that would quickly refreeze the following winter. The presence of these large frozen masses would have a tremendous chilling effect on the climate, particularly during the summer.

FLOODS

Floods—too much water in areas that are not normally under water—are caused by storms driven ashore by hurricanes or coastal storms, torrential or persistent rain, or runoff from rapidly melting snow and ice. Dam failures or natural dams made of debris or ice, may also bring about flooding. On flat terrain, floods usually cover wide areas with slow-moving water; in canyons and valleys, floodwaters flow faster and are potentially more destructive.

285–293 Floods
Flash Flood, Storm-surge Flood

Description: *A body of water or a watercourse that overflows its natural banks or ordinary boundaries.* Some floods are the result of storms or unusually high levels of precipitation, but others are annual and predictable, the result of snowmelt. *Flash floods are sudden rises and falls of streams, usually resulting from brief but intense rainfalls over localized areas. A storm-surge flood is an abnormal rise of sea level along a coast caused by strong onshore winds during a storm or hurricane.*

Environment: Virtually any meteorological event that drops enormous volumes of water in heavy or widespread rainfalls can cause a flood. Some of the rain is absorbed by the ground; most of the rest runs into streams and rivers. The excess above and beyond normal stream flows is floodwater (292, 293).

The nature of a flood is affected by the terrain into which the excess water flows. In flat areas, floods usually inundate wide areas, with slow-moving water (285, 286). Flooding is exacerbated if the soil is saturated from earlier rain, extremely dry with a hard-pan surface, frozen, or paved. Rapidly melting snow cover can add to the runoff from heavy spring rains. Floodwaters on flat terrain can cause tremendous property and crop damage but usually cause little loss of human life. Floods channeled into canyons and valleys are restricted in area (287, 288), but the water flow is faster (292) and potentially more destructive.

Season: Coastal storm-surge floods occur in the fall and winter storm seasons. River flooding is most likely in early spring in the East, late spring in the West. Flash floods strike during the summer months.

Range: Storm-surge floods occur along the Pacific, Atlantic, and Gulf coasts. Melting of the mountain snowpack by unseasonably hot weather is often the cause of flooding in the West and in the intermountain region. In the East, river flooding is likely to result from protracted and widespread rainfall on frozen or snow-covered ground.

Variations: Floods may occur in cities and towns (285, 288), on farmland (286), or in normally dry valleys, arroyos, and washes (290). A flood may be a storm surge driven ashore by a hurricane or coastal storm; water descending from intense local storm clouds; slow accumulation from persistent, steady rain; or runoff from the rapid melting of deep snow cover. Floods can also be released by dam failures or arise upstream behind natural dams of debris or ice jams. Most great floods result from some combination of these causes. Normal seasonal floods include the spring flooding of the Mississippi River bottomlands and of the valleys and canyons of the Colorado River basin (291) and the most famous of river floods, the annual inundation of Egypt's Nile Valley (now a nonevent, owing to the damming of the Nile). Torrential localized downpours from slow-moving or stationary thunderstorms can unleash flash floods, which are often characterized by severe erosion and silt-laden water (290). In densely vegetated locations, flash floods can occur without much silt (292). Unusual amounts of floating debris are characteristic of most floods (289).

Significance: On the average, floods extract an annual toll of 200 lives and $2 billion damage in the United States. Flash floods are especially lethal: Their sudden arrival and swiftly moving water can easily trap and drown unsuspecting victims. The worst flood in North American history was the inundation of Galveston, Texas, by a hurricane in 1900 that drowned at least 6,000 people. The next two deadliest floods, the Johnstown, Pennsylvania, flood in 1889 and the hurricane-related overflow of Lake Okeechobee, Florida, in 1928, each drowned 2,000 or more people; both involved dam and levee failures. Floods also remove enormous amounts of topsoil (300 million tons were washed away in the 1937 Ohio Valley flood), some of which goes into the sea but much of which is often deposited downstream. Hurricanes and tropical storms have also caused severe river flooding, the most extreme case being the deluge from Tropical Storm Agnes in 1972, when the flow of the Susquehanna River approached that of the world's largest river, the Amazon, causing nearly $4 billion damage.

DROUGHT AND RELATED EVENTS

Occasional dry spells occur almost everywhere; abnormal, persistent, and damaging dry weather constitute a drought. In addition to causing shortages in water supply and inflicting damage to crops, droughty conditions can also lead to dust storms and set the stage for severe forest and brush fires.

294–302 Drought

Description: *A sustained and abnormal deficit of precipitation* that disturbs the course of plant growth, causes shortages in the water supply, and results in other disorders in the ecology and economy of a region. The size of the deficit that produces a drought varies with location and season and depends on the normal precipitation, the normal variability of precipitation, the local demand for water, the season, and the duration of the shortage. Droughts are distinct from perpetual dryness—arid or semiarid climate—in areas where over the years the local economy and ecosystem have adjusted to constant and predictable shortages of water. Droughts are also different from dry spells—periods of two weeks or longer without measurable precipitation (0.01″/0.25 mm or more). Dry spells stress lawns and gardens severely but usually have little effect on water supplies, crop production, and wildlife. Most droughts persist for months, or even years, and they cover many thousands of square miles.

Environment: Water shortages result from deficits in the net water budget (precipitation minus evaporation), rather than from shortfalls of precipitation alone; temperature, humidity, and wind are all contributing factors. During a drought, the moisture content of the atmosphere above the drought region actually differs little from what is observed during moist periods. Storms, however, are required for atmospheric moisture to be converted into precipitation; a shortage of storms (or the shift of storm tracks) is the essential cause of drought. Strong upper-level ridges that suppress the growth of thunderstorms during the normally rainy summer season in the southwestern mountains and deserts, the Plains States and provinces, and the Southeast (especially Florida) can cause droughts that are disastrous to summer crops. Even an absence of hurricanes, tropical storms, and other tropical disturbances along the Atlantic and Gulf coasts can result in drought. During the great drought in the northeastern states during 1962–66, coastal cyclones tended to track several miles offshore rather than along their normal path right along the coast; their precipitation fell largely on the ocean. In normal winters, storms strike the entire Pacific Coast from California to Alaska, but drought strikes California when the southern fringe of the storm track diverts north to Washington and British Columbia.

Variations: Winter snow droughts resulting in meager snowpack accumulation create shortages in the water supply.

Significance: Depleted soil moisture during drought can damage vegetation; severe droughts can cause stunted growth or leaf loss in perennial plants (295). Dry soil is less cohesive and less dense than wet soil, and it is therefore more readily lifted by wind; blowing topsoil or dust storms (297) may be seen during droughts in areas where such storms normally do not occur. The levels of reservoirs, where steady depletions are countered by input from streams and rivers, provide graphic evidence of drought-caused water deficits (296). The depth of groundwater supplies, or water table, may drop during drought, and in normally wet areas such as swamps and ponds the water table may drop below ground level, resulting in the characteristic cracked, dried mud (294, 298). Some wet areas, such as the Florida Everglades, have a dry season that brings with it some of the visual characteristics of drought; nonetheless, such dry seasons are a normal and annual occurrence, not a drought.

In areas with permanent drought— areas having arid and semiarid climates—many of the visual signs of drought are present. Cracked mud, for example (299), is common in dry climates, but there it results from wet weather rather than drought. Blowing dust and dust storms are common during dry spells in semiarid climates (302); sandstorms (300) are seen in desert climates. Topsoil particles can be blown great distances, but sand grains are generally much larger and accumulate in sand dunes (301). The ripples on sand dunes form perpendicular to the wind direction (301); the sizes and shapes of entire dunes are strongly affected by the speed and steadiness of direction of the wind.

Comment: Severe droughts in agricultural areas can reduce total North American yields of individual crops by 50% or more, with 10–20% decreases in yearly production of entire crop categories such as grains. Topsoil loss during severe and protracted droughts, such as the seven-year "Dust Bowl" drought of the central United States during the 1930s, can cut production for years after the drought is broken.

The effects of snow shortages are not seen until after the spring melt, when the meteorological drought translates into a water shortage—a delay of several months. Ski areas may lose half their business when the slopes are bare. Open winters on the northern plains, where snow cover normally protects winter wheat and limits the loss of windblown topsoil, can reduce yields the following summer. Loss of the insulating effect of deep snow cover in cold areas may allow the ground to freeze to deeper than normal depths, so underground water pipes freeze.

297, 300, 302   Dust Storms and Sandstorms

Description: *Wind-borne dust or sand lifted high enough* above the ground (eye level or higher) *to reduce horizontal visibility* to ⅝ mile (1 km) or less. The dust or sand is sometimes of local origin, although dust can travel great distances to reduce visibility in locations where normally dust is not a problem. Because of the heavier particles, sandstorms (300) are usually confined to sand-covered terrain and limited to heights within 10′ (3 m) of the ground (as high as 50′/15 m in severe sandstorms). On the other hand, dust may be raised as high as 15,000′ (4,572 m) in severe dust storms. Dust devils, tornadic debris clouds, and microburst dust clouds are generally too brief and localized to be considered dust storms.

Environment: The actual wind speed required to cause a dust storm or sandstorm depends on particle size, soil wetness, and amount of plant cover. Dust and sand may be blown along the ground (300) by winds of 20 mph (32 kph), but such storms generally require winds of 40 mph (64 kph) or stronger (302) to raise the particles to eye level or higher. Dust storms and sand storms originate in the passage of cyclones and gust fronts over any dry soil, in chinook winds, and in winds passing over volcanic ash deposits and very fine-grained "rock flour" deposits.

Season: Late winter and early spring in the Southwest and Plains States; any time of year if winds and topography are correct.

Range: Great Plains and southwestern deserts during the passage of cyclones and cold fronts; dry valleys of western North America; volcanic ash deposits in Alaska and the Pacific Northwest; glacial plains of Alaska and other arctic locations.

Variations: Dust storms that originate in a gust front passing over any dry soil are called by their Sudanese name, *haboob*. Dust storms are colloquially called "black blizzards" and "dusters." Meteorological observers distinguish between storms in which dust is actively raised from the ground by the wind, those in which suspended dust has arrived from elsewhere or from previous high winds, and those in which dust is visible in the distance. In a severe dust storm, visibility decreases to 5/16 miles (0.5 km) or less.

Significance: Among the immediate effects of dust storms are removal of topsoil, fouling of water supplies and machinery with dust and sand particles, and silicosis, a respiratory problem caused by accumulations of particles in the lung. During the last Ice Age, enormous dust storms along the periphery of the continental ice sheet deposited deep layers of rich soil, or *loess*, in the Ohio Valley area. At the same time, powerful chinook winds along the eastern slopes of the Rocky Mountains scoured numerous shallow basins, many of which are now reservoirs. Even now, shallow streams of blowing sand are at work sculpturing bizarre rock forms in desert areas. During severe droughts, such as those during the 1930s and in 1977, dust clouds from the Plains States reduced visibility and caused "black rain" and "brown snow" along the Atlantic seaboard. Beach sand driven by hurricane winds of 100 mph (160 kph) or more have been known sandblast property and people.

303–308 **Forest Fires, Brush Fires**
Wildfires

Description: *Wildfires burn uncontrolled in a forested, wooded, scrubby, or grassy area. Forest fires occur chiefly in old-growth or second-growth forest; brush fires* are those that *occur in scrub areas. Grass (or rangeland) fires burn in grassy plains and prairies.* Wildfire can occur in almost any setting—it is a rapidly spreading and extremely destructive blaze. Large, intense *forest fires* (303, 306) *release enormous quantities of water vapor* (a product of combustion) that may rise to *generate cumuliform clouds* (304). In extreme cases these fire-spawned pyrocumulus clouds may become thunderstorms. Rather than dousing the fire with rain, however, pyrocumulus thunderstorms more often than not exacerbate the problem by starting new fires with their lightning.

Environment: Drought or seasonal dry weather often sets the stage. The supply of moisture to tree and plant roots decreases, and the loss of water by transpiration from leaves increases. As their total moisture content declines, trees and plants become more susceptible to ignition. Dried plant matter on the ground adds to the supply of combustibles. A high wind contributes to the chances of conflagration, which may begin as a result of lightning or spreading sparks from another fire in a nearby locale. In dry summer months, most fires in the West and Alaska are caused by lightning. In coastal California the normal summer desiccation is followed by searing Santa Ana winds from the deserts.

Season: Chiefly during dry spells or droughts during warmer months of the year; autumn in coastal California; winter dry season in the Everglades; spring in deciduous forests of the Northeast; midsummer to late summer in most of the West and Alaska.

Range: Any place with sufficient vegetation to burn.

Variations: Less intense fires are often confined to the forest floor (305, 308), where they burn the highly flammable accumulation of fallen leaves and branches along with the standing trunks of dead trees. Intense fires may spread to the crowns of living trees, especially when the trees are stressed by drought (303). The intense heat of these "crown fires" can send smoke, water vapor, and other byproducts of combustion high in the atmosphere (304, 306). In the most severe fires, the intense heat coupled with the right atmospheric conditions (moderate instability and light winds) can create a fire storm, in which gale-force winds blow toward the core of the fire to feed the heat-driven updraft.

Significance: Forest and brush fires are not, strictly speaking, meteorological phenomena, but they often have meteorological consequences. Fires inject tremendous amounts of microscopic particles of ash and terpenes into the atmosphere. These particles may be considered pollutants by people living downwind from a fire, but they act as condensation nuclei, essential to the generation of clouds and precipitation. Every year an average of 1,000 forest and brush fires burn a million acres in the United States, and a similar number in Canada. Some consider that these fires play a vital role in maintaining healthy forests by removing accumulations of dead material and unhealthy plants and by making space for new growth (307). Nutrients are replaced in poor, thin soils. Fire also opens up a staid forest to a new succession of pioneer seedlings of all different types—contributing to renewal and diversity in the forest. Yet unfortunately, healthy trees and wildlife also perish in fires.

OPTICAL PHENOMENA

Dazzling us in myriad forms and colors, optical phenomena are the visual results of a wide variety of physical conditions in the atmosphere. These phenomena occur through the reflection, refraction, and diffraction of light. Some, such as haloes, twilight, and rainbows, tell us about approaching weather. All impress us with their unique and natural beauty.

328, 359–364 **Aurora Borealis**
Northern Lights

Description: *A luminous glow in the night sky* that may appear *in forms varying from a nearly featureless arc or band to a highly structured and detailed ray or curtain.* The featureless, or quiescent, arcs usually change very little with time; the more structured rays and curtains often evolve rapidly enough to create the impression of motion. The colors range from pale green (in the less active forms) to combinations of red, green, yellow, and violet in the more active displays. Most forms have a fairly sharp lower edge and a more diffuse upper boundary. In the Southern Hemisphere, the aurora is called the aurora australis (or southern lights).

Environment: Auroras occur in the ionosphere at altitudes of 50–600 miles (80–960 km); they are most often seen at high latitudes (the northern tier of states and most of Canada). The light is fluorescent, caused when high-energy electrons emitted by the sun encounter atmospheric atoms and molecules, primarily oxygen and nitrogen. Before reaching the ionosphere, the solar electrons are trapped in the earth's magnetic field and diverted north and south toward the magnetic poles. The stream of electrons reaching the ionosphere is normally most intense at a distance of about 1,500 miles (2,500 km) from the magnetic poles; auroras are most frequent in this zone, called the auroral oval. Extremely energetic electrons discharged by solar flares and other disturbances on the sun penetrate deeper into the magnetic field before being deflected, reaching the ionosphere farther from the earth's geomagnetic poles. In such cases, the auroral oval expands and the aurora is seen at lower latitudes.

Season: Auroras may occur at any time of year, although there is a noticeable preference for the months surrounding the equinoxes: March, April, September, and October. More importantly, auroral occurrence is highly dependent on an approximately 11-year cycle of solar activity known as the sunspot cycle. In the United States most auroras are seen near the peak of the sunspot cycle; sightings are extremely rare during the cycle's minimum. The most recent maximum was in 1989, and subsequent peaks may be expected at successive 11-year intervals.

Range: The aurora borealis is most frequent within the auroral oval, which in North America is a zone crossing northern Alaska, the Yukon and northern Northwest Territories, northern Quebec, and Labrador. In the absence of clouds the aurora may be seen 100 or more nights a year in this zone. The frequency drops to 30 nights per year across much of southern Canada, and to 10 nights per year along the northern tier of states. In the southern United States the aurora appears about once a year. These frequencies assume clear skies; cloudiness may cut the number of visible displays by half or more.

Variations: The most common form of aurora, particularly at high latitudes, is the homogeneous quiescent arc, which may retain an unchanging appearance for hours. If the base of the arc is beneath the horizon, the aurora may simply appear as a diffuse glow along the northern horizon. The appearance of vertical rays and curvature in an arc and an overall brightening (359) are often clues to further and more spectacular development. The rays may extend to great heights (362, 363), and the arc may separate into two or more components (363). Pulses of light may rise along the rays, taking less than a second to reach the top. During very active displays the multiple arcs and rays evolve into curtains (328, 360, 364), whose rapid undulations blur time-exposure photographs. To an observer within the auroral oval, the nearly vertical rays (aligned along the earth's magnetic field) may appear to converge at a point slightly south of the zenith, forming a corona (361). Green and red auroras get their color from oxygen atoms. Violet auroras (363) are relatively rare; the color comes from nitrogen atoms.

Significance: The enormous influxes of electrons associated with auroras cause rapid fluctuations in the earth's magnetic field that in turn may induce significant electrical currents in long conductors such as telephone wires, power lines, and pipelines. Occasionally, power supplies are affected and communications are disrupted. The electrons may further disrupt communications by altering the ionosphere's ability to reflect radio waves at certain frequencies (particularly shortwave frequencies). Auroras have no direct influences on the weather, but some meteorologists think the solar disturbances that cause them may also affect storm development and droughts.

335, 336, 338, 339	**Haloes**

Description: Bright *rings of light in a circle about the sun or (more rarely) the moon.* Some haloes are muted red toward the inside and blue outside; others are colorless.

Environment: Haloes are caused by refracted light passing through ice crystals in cirriform clouds that lie between the observer and the light source (in theory, haloes can be seen around any light source, under the proper conditions). The ice crystals that form the clouds responsible for haloes are usually present high in the troposphere in one or more of the cirriform clouds (335, 336, 338). Ice-crystal formation requires temperatures of about $0°F$ ($-17.8°C$) or lower. Most ice crystals are hexagonal with two flat ends; the relative proportions of length and width can vary greatly. Two crystal types are responsible for most haloes: hexagonal plates, which look like flat, six-sided wafers; and hexagonal columns, which resemble segments of a six-sided pencil. Both shapes have eight surfaces that can refract and reflect light.

As ice crystals fall, those more than a few thousandths of an inch across orient their longest dimensions horizontally. Plates fall flat, and columns fall on one of the six sides. Turbulent winds, especially near the ground, may scatter the crystals into random orientations, diffusing or completely destroying the optical pattern. The most common angle of refraction through a hexagonal shape is approximately $22°$, however, the angle is slightly greater for blue light and less for red light, which is why blue is seen on the outside of a halo and red on the inside.

Season: Any time of year when the requirements of temperature are met.

Range: Anywhere that ice-crystal clouds form; most often observed in the Southwest, where the climate is dry and there is usually no intervening cloud layer.

Variations: The most common ice-crystal optical phenomenon is the 22° halo (335, 336, 339), caused by the refraction of sunlight by hexagonal columns oriented at right angles to the sun in a cirriform cloud between the sun and the observer. The halo's name refers to the size of its radius (the distance from the center to the circumference). Colors are usually weak and show red inside (toward the sun) and blue outside. A much larger 46° halo, created by refraction through one end and one side of hexagonal ice-crystal columns, is possible, but rarely seen. Haloes of other sizes (9°, 18°, 20°, 24°, and 35°; see plate 338) result from refraction of sunlight through the pyramid-shaped ends of some hexagonal ice-crystal columns; they are also quite rare. The moon may also be the light source for a 22° lunar halo.

Significance: The well-known adage about haloes preceding rain or snow by 24 to 48 hours is valid in moister regions subject to the effects of traveling middle-latitude cyclones. Halo-producing cirrus clouds are usually the first clouds to develop from an approaching cyclone. In drier climates, however, cirrus clouds (and haloes) frequently pass without any subsequent precipitation.

337, 339 Parhelia

Description: Parhelia, also called *"sun dogs"* or *"mock suns,"* are *bright spots,* often *tinged with color,* that appear in the sky to the left and right of the sun at a distance of 22°–24° and at the same altitude as the sun. Parhelia usually appear in pairs, but sometimes only one is visible. Like 22° haloes, parhelia may be muted red on the sunward side and blue on the outside.

Environment:	Parhelia are created by sunlight refracted through a 22° angle by horizontally oriented hexagonal ice-crystal plates. Parhelia often accompany 22° haloes (339) and can also be seen in the absence of a complete halo (337).
Season:	All year long if ice-crystal columns are present in the atmosphere; more common in colder months of the year.
Range:	Anywhere cirriform clouds are likely to form; also in cold climates or northern latitudes where ice crystals may form near the ground.
Variations:	The moon is occasionally bright enough to generate observable "moon dogs."

330, 336, 339, 340   Arcs

Description: *A variety of curved arcs of refracted light.* The *upper and lower tangent arcs* (one above and one below the sun) always *appear 22° or more from the sun* and merge with the 22° halo (if present) directly above and beneath the sun (339). The shapes of the arcs change with the elevation of the sun; at high sun angles, the arcs curve toward the sun to form an elliptical circumscribed halo only slightly larger (at its ends) than the 22° halo (336). When the sun is low, the arcs curve away (330); when the lower tangent arc is beneath the horizon it is visible only from airplanes and mountaintops. The *circumzenithal arc* (340), one of the most colorful of the ice-crystal phenomena, is always seen high in the sky. Its curve *is centered on the zenith, and it is seen only when the sun is within 18°–26° of the horizon.*

Environment: The upper and lower tangent arcs are caused by refraction of light above and below the sun through horizontally oriented ice-crystal columns. The circumzenithal arc is caused by refraction through horizontally orient hexagonal plates; the sunlight is refracted through one side and one face of the crystal (rather than two sides, as in parhelia).

Season: All year.

Range: Throughout North America.

339, 342, 345, 346   **Sun Pillars, Parhelic Circle, and Subsun**

Description: *Vertical, horizontal, or disklike reflections of sunlight in the sky.* Reflection phenomena are colorless unless the sun itself is red.

Environment: Unlike parhelia, arcs, and haloes, which are caused by light refracted through ice crystals, sun pillars, parhelic circles, and subsuns are caused by reflection of light. Ice-crystal plates fall flat, and ice-crystal columns fall on one of the six sides; in both, the longest dimensions are oriented horizontally. Horizontally oriented ice-crystal plates (occurring in cirriform clouds or in ice fog near the earth's surface) reflect sunlight into vertical sun pillars (345, 346) above the rising or setting sun. Sun pillars may also be seen below the sun. The subsun, an image of the sun reflected from horizontal ice-crystal plates (342; also seen in plate 262), is seen beneath the horizon from mountains or airplanes. Vertically oriented surfaces, such as the end faces of horizontal ice-crystal columns or the edges of ice-crystal plates, can reflect light in a horizontal band called the parhelic circle (337, 339), which may extend all the way around the sky at the same elevation as the sun. The parhelic circle may combine with a sun pillar to form a cross centered on the sun (337).

Season: All year. Ice fog is common in winter in the far north and occurs during outbreaks of very cold air farther south.

Range: Throughout North America. Ice fog is most common in northern latitudes of arctic and subarctic North America. Ice fog crystals also may form at night in the high and normally cold valleys of the intermountain region.

331, 332 Coronas

Description: *Pastel-hued rings in clouds in the general direction of the sun or moon.* The colors that appear near the sun are often uncomfortably bright; those around the moon are generally easier to observe.

Environment: Coronas are created by the diffraction of light waves passing through relatively thin clouds containing small water droplets of a fairly uniform size. The characteristic droplet sizes and concentrations are most often found in altocumulus and altostratus clouds. Lenticular altocumulus clouds and pileus clouds atop cumulonimbus are common sources of iridescence and coronas.

Season: Any time of year; more common during winter in mountains.

Range: Altocumulus clouds causing iridescence and coronas may occur at any location; lenticular clouds over the western cordillera (and to a lesser extent over the Appalachian Mountains) appear to produce the most frequent displays.

Variations: The diffraction of light through a perfectly uniform cloud layer containing droplets of nearly uniform size produces alternating bluish and reddish rings centered on the sun or moon. Small droplets produce larger ring patterns than do larger droplets; the radius of the innermost ring ranges from less than 1° to about 10° for typical droplet sizes. Distinct colors and nearly circular rings of a corona (332) result from nearly uniform droplet sizes within the cloud layer, while muted colors and irregular ring patterns (331) are produced by droplets of varying sizes.

333, 334 Iridescence

Description: *Colorful bands or patches observed in clouds seen near the sun or moon.* Near the sun the colors may be too bright and intense for comfortable observation.

Environment: As with the corona, the colored bands are due to light being diffracted by water droplets of a fairly uniform size in altostratus and altocumulus clouds, and occasionally in cirrocumulus clouds and pileus clouds atop cumulonimbus.

Season: Any time of year; most frequent in winter over mountains.

Range: Wherever suitable cloud formations occur; virtually everywhere in North America. Most common and frequent over the western cordillera and the Appalachian Mountains.

Variations: Nonuniform droplet sizes within a cloud generate irregular patches of iridescence (334); the grading of droplets to smaller sizes near the edge of a cloud layer may produce a fringe of iridescence (333).

341 Glory

Description: Alternating *reddish and bluish rings surrounding the point directly beneath the sun.* The observer's shadow is often visible at the center of the rings. The rings are similar to those of a corona, and the glory is sometimes called the "anticorona."

Environment: A glory results from sunlight being diffracted and scattered back toward the sun by water droplets in a cloud. Since the sun must be above the horizon, a glory is always seen beneath the horizon. Most glories are observed from airplanes or mountains.

Season: All year.

Range: Anywhere there are water-droplet clouds.

Comments: A glory, or anticorona, is also known among mountaineers as the "Brocken bow," named after a peak in Austria, and the "mountain specter."

343, 344, 355 Crepuscular Rays

Description: *Contrasting, alternating bright and dark rays in the sky* that appear to radiate from the sun. Since the light originates from the distant sun, the rays are nearly parallel, but because of the observer's perspective they appear to diverge. The scattering of sunlight by molecules and particles in the air renders the bright rays visible; the contrast between bright and dark rays is enhanced by haze, dust, or mist. Crepuscular (or "twilight") rays appear after sunset or before sunrise and radiate upward from beneath the horizon (355); the term also refers to rays radiating from the sun at any time of day, usually from behind a cloud (343, 344). The silver lining seen at the edges of cumulus clouds (343) is essentially a narrow band of iridescence; the lack of color is due to the varied and generally large droplet sizes.

Environment: Crepuscular rays require an intermittent obstruction to sunlight, such as scattered or broken clouds or a mountain range.

Season: All year.

Range: Visible everywhere in North America.

Variations: Scattered clouds cast what appear to be dark rays (343), and nearly overcast skies create the impression of bright rays (344). However, the two phenomena are actually the same. Bright rays extending down to an ocean or lake were once referred to as the "sun drawing water," from the mistaken impression that the rays were streamers of mist rising from the water. Rays seen after sunset or before sunrise are caused by shadows cast in the twilight sky by distant mountains or clouds.

329, 347–354   Rainbow

Description: A *multicolored arc or circle, 42° in radius, centered on the point in the sky directly opposite the sun*; seen when falling rain is illuminated by direct sunlight. Sometimes a fainter, secondary rainbow with a 51° radius is also seen.

Environment: The need for falling rain and direct sunlight limits the meteorological conditions under which rainbows can be seen. Portions of the rainbow are above the horizon only when the sun is less than 42° above the horizon—within about three hours of sunrise or sunset. Rainbows chiefly form in association with isolated rainshowers and thunderstorms, or with widespread showers separated by reasonably large clear areas.

Most raindrops are nearly spherical, but drops larger than $\frac{1}{16}''$ (1.6 mm) flatten out on the bottom as they fall; drops smaller than $\frac{1}{16}''$ (1.6 mm) are most effective at making rainbows. Most of the sunlight entering a raindrop passes directly through, but some of the light entering the drop is refracted into its component colors, reflected off the back interior wall of the drop, and then refracted again as it exits the drop. With perfectly spherical drops, reflected light can deviate as much as 42° from the line pointing directly back toward the sun, the light being reflected at all angles up to 42°. Rainbows form at all radial angles up to 42°; at smaller angles, however, the rainbows overlap and their colors wash out. Only the last (and brightest) bow, the one at 42°, survives as a visible rainbow (349). Occasionally several of the outer bows, called supernumerary rainbows, are visible as colored fringes inside the main rainbow (329). Sunlight reflecting twice inside the raindrops is reflected at a 51° angle, producing a larger but fainter secondary rainbow (347, 350).

Season: All year in the tropics and subtropics; chiefly summer elsewhere. Over most of North America, rainshowers are more numerous in the afternoon than in the morning; thus, most rainbows are seen in the east as the sun sinks in the west.

Range: The appropriate showery weather conditions may occur anywhere but are most common in the tropics and subtropics (Hawaii, Florida). Rainbows are most frequent in climates where showers tend to be isolated and least frequent in consistently wet and cloudy climates.

Variations: In the 42° primary rainbow (329, 348, 349), the colors appear in spectral sequence, from violet on the inside to red on the outside. The additional reflection in a secondary rainbow reverses the color sequence; in a secondary bow, red is inside and violet outside. Occasionally supernumerary rainbows are seen outside the secondary rainbow. Neither a single nor double reflection returns any light at angles between 42° and 51°; therefore, the zone between the primary and secondary rainbows is perceptibly darker than the sky elsewhere (347, 350). Sunlight shining through breaks in stratiform clouds usually produces incomplete rainbows; falling snow does not produce a rainbow.

Rainbows may also be caused by spherical water drops from other sources, such as waterfall spray (351–353) and lawn sprinklers. On rare occasions, moonbows (with a very bright full or nearly full moon as the light source) have been sighted in nocturnal rainshowers, primarily in the tropics. Other light sources, such as streetlights, can also produce nighttime rainbows.

Fogbows, also known as white rainbows or cloudbows, may be seen in fog (354) or clouds; they are usually wide, with weak (if any) colors.

Significance: Most showers in North America occur
in the afternoon and are moving to the
east, hence the appearance of a rainbow.
The appearance of a rainbow often
marks the end of a storm.

355, 356   Twilight

Description: *A continuously changing, multicolored glow*
seen in the evening sky after sunset and in
the morning sky before sunrise.

Environment: Twilight is caused by the scattering of
sunlight by molecules in the upper
troposphere or stratosphere when the
sun is beneath the horizon. Morning
twilight begins and evening twilight
ends when the sun is 18° beneath the
observer's horizon. When the sun is
more than 18° below the horizon, the
portions of the atmosphere illuminated
by the sun are so high (50 miles/90 km
and higher) that the scattered light is
not perceptible. Twilight encompasses
the entire sky, but the illuminated
atmospheric layers are lower and denser
in the direction of the sun, and the
resulting glow is brighter.
Light at the blue end of the spectrum is
scattered in all directions much more
widely than is light at the red end.
During the day, rays of sunlight
passing overhead are broadly scattered
toward the observer, thus appearing as
the blue light of the daytime sky.
When the sun is rising or setting, the
blue light is removed from the line of
sight, thus reddening the sun (see the
sun pillars in plates 345 and 346).
The light that reaches the upper
atmosphere from the distant setting sun
is red, but the preferential scattering of
blue light generates a variety of hues,
ranging from red and yellow near the
horizon shortly after sunset and before
sunrise to green, blue, and violet
higher in the sky and during the darker
stages of twilight (356).

Season: Generally, twilight is more prolonged around the time of the summer and winter solstices. At extreme northern latitudes (north of 48°), twilight lasts all night on and around June 22, the date of the summer solstice.

Range: Twilight lasts longer at higher latitudes, where the rising and setting sun approaches the horizon at relatively shallow angles, than in the tropics.

Variations: Distant clouds blocking the illumination of the upper atmosphere can reduce the brightness of twilight; shadows cast by isolated distant clouds may generate crepuscular rays. The brightness and color of twilight are affected by increased particulate matter in the atmosphere. In particular, volcanic haze particles result in brighter reds and purples (355); the unusual twilights may persist for two or three years.

Significance: The appearance of a bright red twilight after sunset indicates the absence of tropospheric clouds for several hundred miles to the west of the observer. A bright twilight before sunrise places the cloud-free area to the east. Since weather systems move generally from west to east across North America, clear weather to the east will be moving away from the observer, likely to be followed by inclement weather (hence the adage, "red sky at morning, sailors take warning; red sky at night, sailor's delight").

357, 358 Green Flash

Description: *An exceedingly brief, brilliant greenish light seen on the horizon immediately after sunset or immediately before sunrise.* To the naked eye, the green flash is so small that it often appears as a point of light.

Environment: The green flash is produced by atmospheric refraction of light, which causes objects near the horizon to appear slightly higher in the sky than they really are. Refraction is stronger for blue and green light than it is for yellow and red light, so the blue light of the sun is lifted somewhat more than the red light. This weak prismatic effect of the atmosphere results in a bluish or greenish fringe on the upper edge of the sun (357). The green flash is visible only when the rest of the sun is below the horizon, leaving the greenish fringe (358). Ripples and undulations on one or more layers in the atmosphere produce notches and lines on the solar disk (357), which sometimes aids in producing the green flash by separating the green fringe from the rest of the solar image.

The green flash requires a flat, low horizon and a cloudless and relatively haze-free line of sight in the direction of the setting sun.

Season: All year.

Range: Most often seen from airplanes and on the ocean—both situations in which the horizon and haze requirements are frequently satisfied.

Comments: Observers looking for the green flash should be careful not to look directly at the sun during the moments before sunset. There is potential for eye damage; what is more, the bright red image of the sun can, after the sun sets, leave a green afterimage in the eye that might be mistaken for the green flash.

OBSTRUCTIONS TO VISION

Visibility below about a half mile (1 km) is characteristic of fog, a term used to describe a cloud that has reached the ground. Haze, made chiefly of water vapor, also contains dust particles and creates a less intense obstruction to vision than fog. Smog is a combination of smoke and fog; it is partially from air pollution, which is a man-made condition caused by the burning of fossil fuels and emission of harmful chemical pollutants into the atmosphere. Volcanic eruptions also lower visibility by releasing volcanic material into the atmosphere and cloud cover.

309–317 Fog

Description: *Fog is a reduction in visibility due to the atmosphere at the ground holding the maximum amount of water vapor that it can contain at that temperature.* Fog is essentially a cloud with its base at the ground. Most of the time, fog is composed of water droplets. When temperatures are much below freezing, ice crystals may form in sufficient quantity to make ice fog. Typically, fog has a gray to somewhat light-blue appearance. *Haze, a natural phenomenon that causes a partial reduction in visibility, may occur before or after fog* has formed. Strictly speaking, *a mixture of fog with smoke is smog.* The term smog is often incorrectly applied to a combination of haze and smoke.

Environment: Fog forms in two basic ways: when moist air cools to its saturation point (dew point), and when moisture is added to the air until it reaches saturation at the ground.
Weather patterns that produce different types of fog are quite varied. *Radiation fog occurs when air cools at night without*

much wind or change in air mass, until the air reaches saturation. Often, the air dries out again in the morning as the temperature increases and humidity is lowered to below saturation. Cumulus humilis clouds may form after a nearly overcast fog and/or stratus cloud cover has lifted (see plates 84–87 in cumulus clouds section). Radiation fog usually occurs at night and in the morning. *Advection fog occurs when cooler air blows into a region while the moisture content of the air is not reduced enough to avoid saturation.* When a large-scale weather system moves across an area that is already rather moist near the ground, falling rain or snow may lower the surface temperature to the dew point, resulting in saturation. Coastal advection fog occurs near colder oceans, such as along the Pacific coast of the United States and Canada, where cool and moist oceanic air moves inland at night to form fog that dissipates somewhat into haze during the warmest part of the day. A small-scale type of advection fog may occur when showers or thunderstorms cool the air near the ground while the moisture content of the air stays about the same; this situation is best seen after a storm has dissipated or moved away. Usually occurring over the northern oceans, and not often encountered on land, advection fog forms when cold air blows across a warm ocean with very moist air near the surface; because air at colder temperatures has less capacity to hold water, fog occurs easily in the colder air.

Upslope fog occurs where there are large changes in elevation, and colder air is lifted as it moves along the ground toward higher altitudes. Along the eastern slopes of the Rocky Mountains of the United States and Canada, cold air from anticyclones often blows from the east to northeast after the passage of a cold front. As the air subsequently moves to the west, it rises and cools further, finally reaching saturation at the ground to form fog, as well as clouds that are deep enough to produce rain or snow.

Season: Radiation fog occurs mainly during the night and morning hours at any time of the year; upslope and advection fog occur mostly during cold months.

Range: Radiation fog is more common in humid regions. Advection fog accompanies large-scale rain and snow systems throughout the continent, and is found with smaller storms in humid regions; along the West Coast, advection fog is seen at any time of the year; it occurs over the northern oceans most often in winter. Upslope fog is most common on the eastern slope of the Rockies during the colder months.

Variations: In plate 311, advancing advection fog is associated with cooling produced by the parent cloud and the precipitation directly above it. Another form of advection fog, illustrated in plate 315, is due to a large-scale weather system that has brought more moist air into a region where the air is near freezing because of snow remaining on the ground. Coastal advection fog is shown in plates 313, 314, and 317. The air touching the ocean is cool or cold, saturated, and very stable; when fog occurs along the Pacific coast, warmer air is often on top of cold surface air. Plates 310 and 312 show valleys filled with fog; plate 10 in the atmosphere section is a satellite photo of California's enormous Central Valley, filled to overflowing with dense fog. It may be that traveling weather systems brought moisture into these valleys, which had already cooled by radiation, and that further radiation at night intensified the situation, as may happen when airflow in a valley is weak for several days. Radiation fog appears to be the main factor in the conditions illustrated in plates 309 and 316, since the sky above the fog is brighter,

indicating that the densest fog is at
the surface, where the temperature is
the lowest.

Significance: Radiation fog is generally a diurnal
occurrence, that is, a weather
phenomenon that follows a daily cycle.
If the sky is mainly clear in the
afternoon, becomes foggy at night,
and begins to clear in the morning,
radiation from the earth's surface is the
main cause of fog. If there is very cold
water nearby, a daily cycle may also be
followed, but the fog will remain just
offshore over the water and move back
in again at the end of the day. When a
large-scale weather system begins to
move into an area before any warming
winds have started, or rain starts to fall
into cold air, fog will be widespread
and may be visible day or night. If
colder, moist air begins to blow toward
higher ground, such as along the Rocky
Mountains, clouds and fog may form in
a north–south band at the same
elevation range for hundreds of miles.

318–326   Air Pollution and Haze

Description: *Haze is a natural phenomenon that reduces
visibility,* often *during periods of medium
to high relative humidity. Air pollution,* on
the other hand, *is man-made.* The two
often occur together over the United
States and Canada and are thus
indistinguishable. *Smog is a combination
of smoke and fog* indicating that high
humidity is present, although it is haze
rather than fog that is usually
responsible for reducing the visibility.
Haze usually appears as a blue to white
veil against buildings, trees, mountains,
or hills. Air pollution and smog may
appear yellow or brown, but they are
often hard to distinguish from haze.
Not every reduction in visibility is
necessarily man-made. Natural haze is
known to have been present centuries

ago in humid regions of the eastern states, notably in the Great Smoky Mountains. Haze was also mentioned in early records of the drier Los Angeles basin.

Environment: Any reduction in visibility is caused by the concentration of dust, moisture, or pollution particles in the atmosphere. The typical situation that brings this on is an inversion in temperature (warmer air above colder air) near the ground, which can occur in several different ways. Cool air can blow onto the land from a nearby lake, river, or ocean, with warmer air above. Cooler air can drain into a valley or basin during the night and may not warm fast enough the following morning to become as warm as the air above. If the ground is covered with snow or ice, it will keep the surface temperature colder than the air above. An inversion of temperature also occurs when large-scale weather patterns produce anticyclones with large areas of colder air near the ground that underlie warmer, sinking air. Many cities along the west coast of the United States have serious air-quality problems owing to a combination of a source of cool and moist air from the adjacent ocean, the drainage of cooler air from nearby mountains, and the sinking of warm air that often takes place overhead in the Pacific anticyclone.

Season: Air pollution and haze can occur at any time of the year; they are most prevalent in summer along the west coast of the United States and appear mostly in winter in the rest of the United States and Canada.

Range: Along the west coast of the United States, where very warm air lies over an ocean-cooled surface layer, the worst haze and air pollution tend to occur during the summer. In most of the rest of the United States and Canada, visibility reductions occur more commonly in winter, especially in the

Rocky Mountain states, where mountains trap colder air near the ground and the humidity is often fairly high.

Variations: Plates 324 and 325, photographs shot at different times from the same location in Washington, DC, offer dramatic evidence of the difference in visibility with and without air pollution. The direct sources of air pollution are shown by the plumes in plates 319 and 320. Plate 322, an aerial view, shows how widespread the sources of pollution can be. In this case, the air is cold and the ground is covered with snow, so the plumes are at least partly composed of water; nevertheless, an artificial amount of vapor injected into the air is also a form of pollution. Plate 318, a view of Chicago, reveals the unnatural color of the atmosphere caused by air pollution, seen here as an orange tint on the buildings. Plate 326 shows both dust and haze in Oklahoma at sunset, when the nature of particles in the sky is often clearly visible. In western states, such as Colorado, where the humidity is not always high, sunlight interacting with vehicle exhaust produces a "brown cloud" during the daytime, when conditions are such that sinking air aloft and stagnant air coexist for several days very near the ground (321, 323).

Significance: Haze is a natural condition in many locations and seasons, and is not a significant problem unless the numerous particles cause allergic reactions. Air pollution, however, a man-made condition resulting from a wide variety of sources, should be recognized, measured, and managed as a hazard to human health, plants, animals, and physical structures. In the last decade, in cities and in such places as the Grand Canyon, visibility has been noticeably reduced by man-made particles polluting the air. Keep in mind, though, that not all reduced visibility is due to man-made air pollution; cooler air overlain by warmer air may cause reduced visibility just as effectively, if not as detrimentally.

327 Volcanic Clouds

Description: In plate 327, *an erupting volcano injects ash into the atmosphere;* the layer looks like a stratus cloud at the height where the volcano is active. Other thin stratiform clouds are scattered in the distance at left.

Environment: First, the volcano in plate 327 gave an initial upward push, as shown by the small vertical bubbles above it. However, the volcanic cloud did not reach far above the mountaintop, because the eruption was not very strong at this time and the temperature change with height is rather stable, prohibiting much vertical cloud growth. The cloud was subsequently blown downwind horizontally in a rather thin layer. This stage looks so much like an altostratus cloud that it is hard to differentiate between them from a great distance except by color, since a volcanic cloud is darker than water vapor clouds.

Season: Volcanic clouds are totally dependant on volcanic activity, which has no seasonal variations.

Range: In North America, volcanic-ash clouds typically occur from Alaska southward to California. Volcanic clouds can extend hundreds of miles from their source after a period of hours or days, and flow with the wind at the injected level. In several episodes that have taken place in the last hundred years, volcanic material has been injected into high levels of the atmosphere—above 60,000' (18,000 m)—where it can linger for months or years. The presence of volcanic ash can cause more colorful sunsets, often including enhanced purple or rose hues (see plate 355 in optical phenomena section), that may be visible around much of the world.

Variations: The kind of cloud that results from an eruption depends on the amount and type of volcanic material that is injected into the atmosphere, the height it reaches, and the time it takes to get there. A heavy mass will not travel far, but material of a fine consistency, especially of sulfur dioxide that lingers after the ejected particles have fallen out, can become widely dispersed before it falls or precipitation brings it down to the ground. Some sort of volcanic activity occurs somewhere on the earth every day, but only the largest events—for example, those that are large enough for the news media to report—are likely to produce significant clouds more than a few tens of miles away.

Significance: Volcanic material can reduce the temperature in the immediate area by up to 20°F. Large eruptions that eject volcanic material into the upper atmosphere can reduce temperatures by a few tenths of a degree over a large area of the earth for up to several years. Large amounts of volcanic ash near eruptions have been responsible in recent years for near crashes of several jets in the United States and Canada, especially when the volcanic cloud was imbedded in other heavy cloud clover. The fall of material from Mount Saint Helens also caused damage on the ground to engines and other equipment that draw in large amounts of air. The influence of volcanic ejections on the local area is important for a relatively short period and small distance, but the ejected material may have a more subtle impact that can last for years. For instance, while volcanic ash can initially destroy crops, it can, over long periods of time, enrich and fertilize the soil.

HISTORICAL ACCOUNTS

Plates 365 through 378, described in the text accounts below, show some of the momentous occasions in the recent meteorological history of North America.

365, 366 Hurricane Damage

View: The Richelieu Apartments, Biloxi, Mississippi, before and after Hurricane Camille, August 1969. Probably the worst hurricane ever to hit our region, Camille had a barometric low of 26.85″ (909 millibars) and gusts of 173 mph (278 kph). The storm killed 144 people.

367 Windstorm Damage, Oregon

View: "Columbus Day Big Blow," Monmouth, Oregon, October 12, 1962. In fall, the Pacific Northwest can be hit by mighty winds, spawned by storms over the Pacific Ocean, that do great damage.

368 Long-term Wind Damage

View: The Tacoma Narrows Bridge ("Galloping Gertie") succumbs, November 7, 1940. Winds inflict long-term wear and tear. Too slender and flexible, "Galloping Gertie" was subject to fits of vertical undulations, and eventually collapsed.

369 Dust Storm, Liberal, Kansas

View: Wind-borne dust, 1950s. Dust storms develop gradually from cyclonic winds

of 45 mph (72 kph) or more. The dust
rises to 12,000' (3,658 m) or higher;
visibility can drop to 3/4 mile (1.2 km).

370 Effects of Drought

View: Colorado cropland in the Dust Bowl
years. The arability of the Great Plains
was vastly enhanced by man-made means
early in the century, but drought in the
1930s cruelly upset the trend. By 1936,
the Dust Bowl covered 50 million acres.

371 Surging Floodwaters

View: The James River near Richmond,
Virginia, August 20, 1969. In 8 hours,
Hurricane Camille dumped 25" (64 cm)
of rain in some areas; in one place the
total reached 31" (79 cm)—the most
believed possible.

372 Flood Damage

View: Storm-related flooding, Yuba City,
California, December 1955. The
rampaging Feather and Yuba rivers
inflicted heavy damage; the floods
resulted from cyclonic storms that
released record rainfall and killed
74 people.

373 Heavy Snowfall, Pompey, New York

View: A 25' (8-m) snowdrift along a highway
in upstate New York, March 10, 1960.
Inland from the Great Lakes, snowfall
from successive storms can reach
great depths.

374 Ice-Storm Damage

View: Severe ice storm, January 10, 1949,
       Bolivar, Mississippi. The storm covered
       a large strip of Mississippi, dropping
       2.6″ (67 mm) of rain in 24 hours while
       temperatures hovered near freezing.
       Wires and trees were covered with a
       thick, icy coating.

375 Heavy Snowfall, New York City

View: Madison Avenue and 50th Street, the
       Great Blizzard of 1888 (March 12–14).
       Considered the most famous storm in the
       Northeast in a century, this blizzard
       dumped 42″ (107 cm) of snow in areas.

376 Snow Rollers, Eden Prairie, Minnesota

View: Snow rollers dot an open field of snow,
       December 26, 1970. One day after 6″
       (15 cm) of very light snow fell, freezing
       40 mph (64 kph) winds rolled the
       snow into both small balls and larger
       cylinders. There are records of snow
       rollers 4′ (1 m) long and 7′ (2 m)
       in circumference.

377 Severe Floodwaters, Cambridge, Ontario

View: Constable Jack Shuttleworth surveys
       a street flooded by the Grand River.
       A flash flood roared through in May
       1974, the result of heavy rains from a
       large, slow-moving front.

378 Severe Floodwaters, Fort Scott, Kansas

View: Where next? Kansas cows contemplate
their fate, October 5, 1986. Heavy rains
in southeastern Kansas lasted a week;
severe flooding resulted from the
combination of a nearly stationary front
and the remnants of Hurricane Paine.

Part IV
Appendices

WEATHER MAPS

The weather map is to the forecaster what the nautical chart is to the navigator and the star chart is to the astronomer: It serves as a guide to his or her daily work. George Stewart, the author of the fascinating meteorological novel *Storm,* has described the weather forecaster's map as "simple, beautiful, and inspiring" to its creator.

Unlike most other charts and maps, which are relatively fixed for our time, the weather map changes daily, even hourly. A cold wave on the Canadian border one morning may be deep in the heart of Texas the next. A storm center on the Pacific Coast at midnight may be on the crest of the Rocky Mountains 24 hours later. The temperature at Chicago may drop as much as 20 or 30 degrees overnight, while it rises the same amount at Pittsburgh. A northeaster in New England may cover the bare ground overnight with a mantle of snow a foot deep. On the weather map, change is the only constant.

Origins of the Weather Map:

The creation of the first useful weather maps awaited the invention of the telegraph. Attempts had been made in the late 1840s and 1850s to assemble weather reports from distant places, but the restricted coverage and lack of knowledge about weather movement limited the reports' usefulness. The exigencies attending the Civil War postponed the creation of a North American daily weather map until the early 1870s. Originally founded as the U.S. Army Signal Service and later renamed the U.S. Weather Bureau, today's National Weather Service (with the cooperation of the Canadian Weather Service), transmits local weather-station reports to Washington, DC, where surface and upper-air maps are prepared for the many specialized tasks of modern weather forecasting. The weather data is entered on the map drawn by a scanner, and the forecasts are prepared. In addition to the preparation of surface and upper-air maps, radar analyses of precipitation for different sections of the country are composed and satellite images analyzed for cloud cover, snow extent, and global temperatures.

Recent Developments: In 1989, the National Weather Service launched a $3 billion plan to modernize the gathering of weather data and the distribution of forecasts and warnings. One thousand weather stations will automatically take observations and transmit them electronically to forecast offices throughout the country. This will greatly reduce the number of forecasting offices. A new radar system that uses Doppler techniques, which can detect local wind variations on airport runways, will be installed. Five new satellites will be added to give global coverage of cloud cover, storm location, and global temperature distribution. High-speed communication methods will replace obsolete teletype machines. These improvements should be made by the mid-1990s.

Daily Weather Maps: The principal maps distributed to the general public are known as the Daily

Weather Maps, Weekly Series. This once-a-week, eight-page publication contains four maps for each of the seven days of the week. On the front page is a description of the formats of the four daily charts.

The largest of the maps is the Surface-Weather Map, showing local weather for about 350 major stations in the United States and Canada, with an analysis of the data for 7:00 A.M., EST. The positions of the high- and low-pressure systems are indicated, and the trend of barometric pressure is plotted. A chain of arrows maps the tracks of well-defined low-pressure systems; white crosses in black squares mark the centers where low pressure existed 6, 12, and 18 hours earlier. Dark shading indicates areas of precipitation. The reports printed in this map represent only a fraction of the stations upon which the overall analyses are based.

A second map, the 500-millibar Height Contours Chart, is an upper-air map depicting the pattern of atmospheric pressure at the 500-millibar pressure height (approximately 17,500 feet/ 5,334 m) and the existing temperatures at that level. These observations are taken at 7:00 A.M., EST. As in a topographic map, contours of the height of the 500-millibar level are shown as continuous lines labeled in dekameters (tens of meters) above sea level. Greater heights of the 500-millibar level indicate high-pressure systems at altitudes of 17,000 to 18,000 feet (5,181 to 5,486 m); lower altitudes indicate low-pressure areas. Isotherms (lines of constant temperature) are indicated by broken lines, giving temperatures in degrees Celsius. Arrows indicate the wind direction and speed at the 500-millibar level. Barbs and pennants show wind speed: Each full barb represents 10 knots, each half-barb 5 knots, and each pennant 50 knots.

February 21, 1991 at 7:00 A.M., E.S.T.

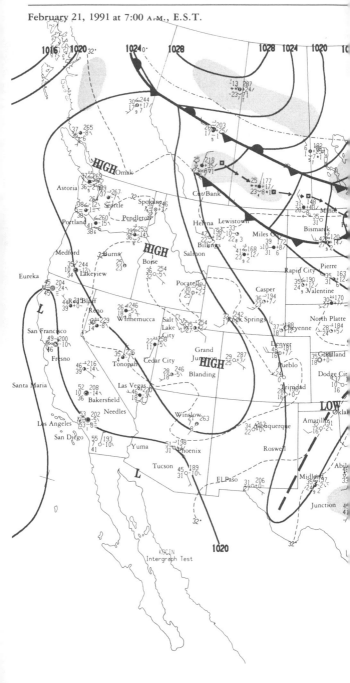

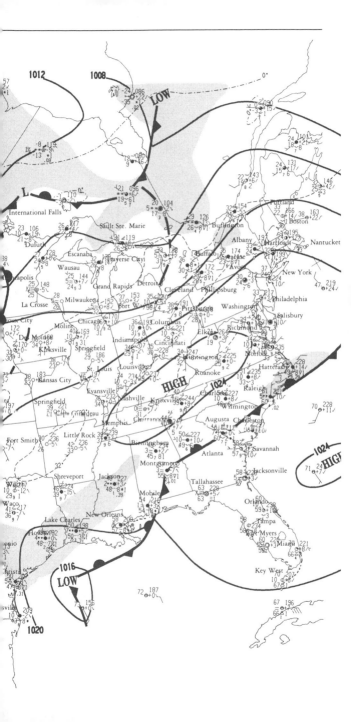

Weather Maps, February 21, 1991

500-Millibar Height Contours at 7:00 A.M., E.S.T.

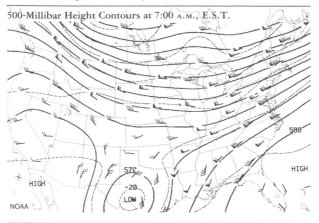

Precipitation Areas and Amounts.

Highest and Lowest Temperatures

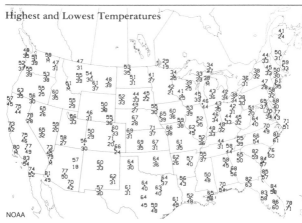

The Highest- and Lowest-Temperature Chart presents the maximum and minimum temperatures at a number of stations in the United States. The maximum represents the 12-hour period ending at 7:00 P.M., EST, of the previous day; the minimum represents the 12-hour period ending at 7:00 A.M. of the current day. (The names of the selected reporting stations are printed on the Surface-Weather Map.) The maximum temperature is given above the station location and the minimum temperature below.

Finally, the Precipitation Areas and Amounts Chart shows, through shading, areas that experienced precipitation during the 24 hours ending at 7:00 A.M., EST, giving amounts to the nearest 0.01 inch (0.03 cm). Incomplete totals are underlined, and "T" indicates a trace of precipitation. Broken lines give the depth of snow on the ground in inches for the period ending at 7:00 A.M., EST.

Making the Synoptic Weather Map: Meteorologists call weather maps *synoptic charts;* synoptic means presenting a broad overview of existing conditions. In meteorological parlance, the word has come to mean synchronous as well, because, according to most schools of meteorology, only those weather observations that are made at the same time under standard conditions lend themselves to rational analysis for the purpose of preparing forecasts. Maps are usually prepared every 3 or 6 hours for surface observations, and every 12 hours for upper-air data.

Weather conditions around the world are recorded by observers for various national weather services, and by observers in ships at sea, in commercial aircraft, and at military bases; this information is then transmitted to forecast centers around the globe in accordance with the regulations of the World Meteorological Organization, a branch of the United Nations with headquarters in Geneva, Switzerland. Standard procedures and symbols are employed so that the weather data will be intelligible to all users.

The Station Model: Data observed by individual weather stations are plotted throughout the world in standard symbols in a standard arrangement on a large surface map like the one described above and shown on pages 580–581. The symbols appear on the chart in a station model, or group, around a printed circle (called the station circle) that indicates the station's geographical location (See the symbolic station model and the sample plotted report on page 586.) An abbreviated station model includes sky cover, surface wind, present weather, temperature, dew point, cloud types, pressure, pressure tendency, and accumulated precipitation. Turn to pages 588–591 for a summary of some of the most commonly used weather symbols.

The direction from which the wind is blowing, given in terms of deviation from true north (a true wind), is indicated on the Surface-Weather Map by a shaft drawn in that direction. Wind speed is indicated by a barb-and-pennant system at the end of the shaft: A full barb on the shaft is valued at 10 knots, a half-barb at 5 knots, and a pennant at 50 knots. (To avoid confusion in length with a single full barb, the half-barb appears partway down the shaft.) The wind speed, rounded to the nearest 5 knots, is the sum of the values of the barbs and pennants appearing on a shaft. A circle drawn around the station center indicates calm conditions.

The amount of black filled in the station circle represents the total amount of sky coverage by all cloud layers over the station.

The sea-level-pressure value is shortened for transmission but can be decoded by employing a simple formula. The pressure to the nearest tenth of a millibar is represented by a three-digit figure. To translate this into the actual pressure, simply place a decimal point before the last digit (e.g., 024 becomes 02.4). If the adjusted number is from 0 to 55.9, a 10 is placed before it (e.g., 55.9 becomes 1055.9 millibars). If the adjusted number is from 56.0 to 99.9, a 9 is placed before it (e.g., 99.9 becomes 999.9 millibars).

Barometric-pressure changes for the three hours before the observation time of the weather map are also indicated on the station model. A plus or minus sign before the number tells whether the net change in pressure is an increase or a decrease. The amount of change is given in tenths of millibars, and the pattern of change is shown by the code symbol.

On surface maps prepared by the National Weather Service, temperature and dew point are given in degrees Fahrenheit and appear to the left of the station circle. In other major countries the values are given in degrees Celsius. The symbols for present weather represent forms of precipitation or obstructions to vision, or both. Although there are 100 possible combinations and types of weather symbols, only a few are regularly employed (See pages 588–591 for a sample of these symbols.)

Many cloud types are symbolized in the international code, but the restricted space available around the station circle limits the number of types presented to ten each of low, middle, and high clouds. Symbols for these clouds are placed either directly above or below the station circle.

Station Model

Symbolic Form of Message

IIiii $i_R i_x hVV$ Nddff $1 s_n TTT$

$2 s_n T_d T_d T_d$ 4PPPP 5appp

$6RRRt_R$ $7wwW_1W_2$ $8N_hC_LC_MC_H$

Note: This abridged code shows only data normally plotted on printed maps.

Sample Coded Message

| 72405 | 11212 | 83220 | 10011 | 20000 |
| 40147 | 52028 | 60111 | 77060 | 86792 |

Symbolic Station Model

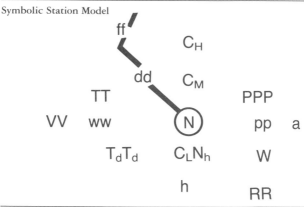

Sample Plotted Report

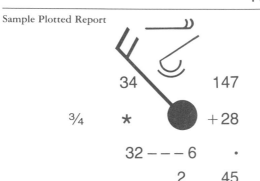

Station Model Symbols and Explanation (sample plotted report)

C_H Cloud type (high cirrus)

N   Extent of cloud cover (sky completely covered)

ff   Wind speed, in knots (18–22 knots)

dd   Direction from which wind is blowing ($32 = 320° =$ northwest)

TT   Current air temperature, in tenths of degrees celcius ($011 = 1.1°C/34°F$)

VV   Visibility, in miles and fractions ($12 = ^{12}/_{16} = ^{3}/_{4}$ miles)

ww   Present weather (slight intermittent fall of snowflakes)

T_dT_d Temperature of the dew point, in tenths of °C ($000 = 0.0°C$ or $32°F$)

C_L Cloud type (low fractostratus and/or fractocumulus)

h   Height of base of lowest cloud ($2 = 300$–600 feet)

N_h Fraction of sky covered by low or middle cloud ($6 = 7$ or 8 tenths)

C_M Cloud type ($9 =$ middle altocumulus)

PPP   Barometric pressure, in tenths of millibars at sea level ($0147 = 1014.7$ mb)

PP   Pressure change in preceding 3 hours ($+28 = +2.8$ mb)

a   Barometric tendency in preceding 3 hours ($2 =$ rising steadily or unsteadily)

W   Past weather (rain)

RR   Amount of precipitation during the past six hours ($011 = 0.45$ inches)

WW Present
Weather:

,	Intermittent drizzle (not freezing), slight at time of observation
• •	Continuous rain (not freezing), slight at time of observation
✳	Intermittent fall of snowflakes, slight at time of observation
∇̇	Slight rain shower
∇̇	Slight snow shower
⌀∿	Slight freezing rain
△̇	Ice pellets or sleet
⌐⌒⌒	Visibility reduced by smoke
∞	Haze
⌒S⤳	Slight or moderate dust storm or sandstorm, no appreciable change during past hour
)(	Funnel cloud(s) within sight at time of observation
⟨↘	Lightning visible, no thunder heard
⎹⎡↘⎤	Thunderstorm, but no precipitation at the station
•/✳ ⎹⎰↘	Heavy thunderstorm, without hail, but with rain and/or snow at time of observation

ff Wind Speed:	*knots*	*miles per hour*
◎	Calm	Calm
——	1–2	1–2
⊥——	3–7	3–8
＼——	8–12	9–14
＼⊥——	13–17	15–20
＼⊥——	18–22	21–25
＼⊥⊥—	23–27	26–31
＼⊥⊥—	28–32	32–37
＼⊥⊥⊥	33–37	38–43
＼⊥⊥⊥	38–42	44–49
＼⊥⊥⊥⊥	43–47	50–54
◣——	48–52	55–60
◣⊥—	53–57	61–66
◣⊥—	58–62	67–71
◣⊥⊥	63–67	72–77
◣⊥⊥	68–72	78–83
◣⊥⊥⊥	73–77	84–89
◣◣⊥	103–107	119–123

Weather Map Symbols

N Sky Cover:

○	No clouds
◔	One-tenth or less, but not zero
◕	Two-tenths to three-tenths
◑	Four-tenths
◑	Five-tenths
◑	Six-tenths
◕	Seven-tenths to eight-tenths
◑	Nine-tenths or overcast with openings
●	Completely overcast (ten-tenths)
⊗	Sky obscured

C_L Low Clouds:

⌒	Cu of fair weather, little vertical development, and seemingly flattened
⌣	Sc not formed by spreading out of Cu
—	St or Fs, but no Fs of bad weather
⋈	Cb having a clearly fibrous (cirriform) top, often anvil-shaped, with or without Cu, Sc, St, or scud

Assembling the Weather observers in the United States,
Data: Canada, and around the world follow
similar procedures for assembling the
weather map. The instructions and
regulations are set by the World
Meteorological Organization.
Weather stations simultaneously
observe and report weather four times a
day, every day. By international
agreement these synoptic observations
take place at 0000 Greenwich mean
time (GMT) (also called Universal time)
and at six-hour intervals thereafter.
Synoptic weather maps are prepared by
the national weather services at one or
more of these times.

Weather The U.S. National Weather Service and
Observers: the Canadian Weather Service rely on a
wide variety of stations and observing
systems. Surface-weather conditions are
observed and reported at 1,200 land
stations, of which about 350 are
operated by weather-service personnel.
Other stations are operated by
government agencies such as the
Federal Aviation Authority or by
contract companies or private citizens.
Observations over the oceans are made
by volunteer observers and transmitted
by radio from more than 2,500 ships.
The weather services also benefit from
the 15,000 cooperative weather stations
that provide daily precipitation totals
and temperature extremes for climatic,
hydrological, agricultural, and other
service programs.
Temperature, moisture, and pressure
profiles from the surface to about
100,000 feet are determined by a
package of small electronic instruments
called a radiosonde, which is carried
aloft by balloons. Wind speed and
direction are determined by tracking
the radiosonde with special radar.
Information from the layer between
100,000 and 300,000 feet is obtained
by meteorological rockets, although
these findings are more useful for

research than for daily operational service programs.

Hundreds of aircraft transmit weather reports. Satellites are used to monitor weather conditions at and above the surface of the earth. Radar is used to provide information on the type, intensity, and movement of areas of precipitation, severe thunderstorms, tornadoes, and hurricanes.

Making an Observation: Armed with a notebook and pencil, a weather observer goes first to the instrument shelter, which is located over a grassy plot to reduce radiation and reflection. After reading the dry-bulb and wet-bulb thermometers of the psychrometer to obtain humidity data, he or she takes a reading of the maximum-minimum thermometer to obtain the range of temperature over the past 24 hours. Next a stick measurement is made of the amount of precipitation in the rain gauge. (At major stations, some of these readings are registered by electronic sensors.) Next the observer looks at the sky, noting the different types of clouds, their direction of movement, their approximate height, and the amount of the sky they cover. At larger stations, the observer will either estimate or measure the height of the cloud levels, using a ceiling balloon that disappears into the lowest cloud level, or at night using a ceiling projector that projects a vertical light beam to the lowest cloud level.

Coding and Transmission: The synoptic observation is coded for transmission and the barometer reading is converted to an equivalent sea-level pressure so that it may be compared with other observations. The observer then composes a message made up of five to eight groups of five numerals each, putting the message in an international code that is understood by meteorologists around the world.

The synoptic report is considered old after one hour and outlives its immediate usefulness after six hours. Thus, the coded report must be transmitted as rapidly as possible after it has been prepared. Most messages are transmitted by high-speed teletype, though some remote locations must use radio or microwave phone. A collection center receives all regional reports and forwards them to a central collection agency, whose computer system automatically plots the reports in their proper places on the daily weather map.

Analyzing the Map: After about an hour and a quarter, the forecaster receives the map. Many forecasters like to draw their own isobars over the computer map, joining all areas of equal pressure. (In the United States, isobars are drawn by computer in Washington, DC.) As the pressure patterns take shape, areas of high and low pressure are shown by the closed isobar curves. When the general pattern becomes apparent, the forecaster looks for fronts (boundaries of various air masses seen on the map), guided by satellite photographs of their accompanying clouds. Highs and lows are appropriately labeled, isobars numbered, and the fronts indicated by symbolic lines. In order to finish the map, precipitation areas must be shaded in, large symbols marking types of weather must be made more prominent, and the names of air masses must be labeled.

Map analysis now takes up to an hour. The map corresponding to observations taken at 5:00 A.M. is a working chart by 7:15 A.M. In the next few years, the United States will work on an interactive computer system that will allow observers to retrieve the data themselves, do their own analysis, and turn data gathered at 5:00 A.M. into a working chart within only 15 minutes!

BECOMING AN AMATEUR WEATHER WATCHER

Nature has put a wonderful weather laboratory as near as your door or window. In any 24-hour period a myriad of sights pass overhead in countless variety, presenting more than enough weather events to enthrall any nature lover.

At first all you need are the color photographs and descriptions of the major cloud types in this book or in any good cloud chart. You will come to associate particular types of clouds with corresponding weather events.

Weather Observing: The first thing to remember is that weather observing is a scientific undertaking: You must be systematic and careful in making each of your observations. Take at least two major readings of the sky each day, preferably soon after sunrise and just before sunset. The National Weather Service makes major observations at 7:00 A.M. and 7:00 P.M., eastern standard time. If you do the same, it will be easy to compare your own readings with the official ones published daily in local newspapers or announced over radio and television. If possible, supplement your twice-daily observations with an additional one at 3:00 P.M. local time, when the sun is usually strongest and the day's temperature is near maximum.

Be sure to record your figures as soon as you determine them; trusting your memory is not a scientific way to observe the weather. A sample recording form to help you get started is shown on page 599. As your interests and knowledge expand, you will be able to amend the form accordingly.

If you miss an observation, do not make one up. You may be able to get substitute information from the newspaper or radio to complete your records. Many radio stations broadcast temperature, humidity, wind, and barometric data as part of hourly news programs, and in large cities this information, plus the forecast, can be secured over the phone. In many places the U.S. and Canadian weather services operate special radio stations that broadcast continuous weather information 24 hours a day. The Weather Channel on cable television also broadcasts around the clock, giving both local and national information.

Be sure to put all your substitute observations in a form that distinguishes them from those that are authentic. For example, if you record your observations on paper, use a colored pencil to denote the substitutions; if you use a computer, underline or boldface these entries.

Getting Started: To begin, you will need to assemble some equipment with which to measure local atmospheric phenomena. The information in this essay will help you learn which devices to use and how they operate. After that you will find some hints about daily observations. You also might consider building a small shelter (with an interior of at least 15 inches/ 38.1 cm square) to protect your various weather instruments. Louvered sides, perhaps made from old window shutters, will permit proper air ventilation, and a double roof, with several inches (about 7 cm) of air space between surfaces, will minimize the sun's direct heat. Bore holes through the floor of the shelter to permit any windblown rain to drip to the ground and put the door on the north side, with the roof sloping slightly to the south. To finish, paint the entire box white. This shelter will provide good ventilation, exposure, and protection for your equipment.

Measuring Temperature: We are all conscious of air temperature because it affects our immediate comfort, but we cannot sense the exact temperature without an instrument. A thermometer measures the heat content of a body such as the atmosphere by the expansion or contraction of a fluid—mercury or alcohol. It is the first instrument you should acquire in preparing an amateur weather station. The Fahrenheit scale is employed in the United States for surface observations; it reads 32° at freezing and 212° at boiling (at sea level). The Celsius scale, reading 0° to 100° for the same span, is preferred in most other nations and Canada. It is universally employed in upper-air meteorological work and in laboratory measurements. A thermometer with dual gradations—in both Fahrenheit and Celsius—is recommended because it will familiarize you with the equivalents.

A thermometer should never be exposed to the direct rays of the sun or to nearby surfaces that can reflect or radiate heat. Mount your thermometer in the shelter you have built, or place it 3–4 inches (7.6–10.2 cm) away from a wall, providing plenty of free space around the bulb to ensure adequate air circulation, and rig a "roof" or sloping board over it to protect it from rain and snow. The best exposure in the Northern Hemisphere is on a north-facing wall about 6 feet (1.8 m) above the ground.

thermometer with dual gradations

Filling In Your Weather Log

You should record weather observations at least twice a day, in the morning and in the evening.

Sky: First, look at the sky and enter the percentage of cloud coverage, using the symbols shown below.

◯   Clear; up to 1/10 cloud coverage

◑   Scattered; 2/10 to 5/10 cloud coverage

◍   Broken; 5/10 to 9/10 cloud coverage

⊕   Overcast; over 9/10 cloud coverage

If some form of precipitation is occurring, enter one of the following letters instead of a cloud-coverage symbol:

R	Rain	T	Thunderstorm
S	Snow	Z	Freezing rain
E	Sleet	L	Drizzle
A	Hail	F	Fog

Temperature: Next, fill in figures taken from your weather instruments for temperature, relative humidity, air pressure, wind speed and direction, and amount of precipitation. For temperature figures, record the thermometer readings to the nearest whole degree. At the morning observation, record both the current reading from the regular thermometer under "A.M." and the minimum reading from the maximum-minimum thermometer under "Min.," remembering to reset the thermometer. At the evening observation, record the current reading from the regular thermometer under "P.M." and the maximum reading from the maximum-minimum thermometer under "Max.," again being sure to reset the thermometer. Then add the maximum

Daily Weather Log

Day	Sky		Temperature					Humidity	Barometer		Wind		Precipitation		Remarks
	AM	PM	AM	PM	MAX.	MIN.	MEAN	AM	PM	AM	PM	AM	PM	AM	PM	
1																
2																
3																
4																
5																
6																
7																
8																
9																
10																

and minimum figures, divide by two, and enter the result under "Mean."

Humidity: Relative-humidity figures should be entered as whole percentages, calculated by comparing wet- and dry-bulb readings from your hygrometer or psychrometer to standard tables.

Barometer: Barometer readings, indicating air pressure, should be entered to 0.01 inch (0.03 cm) or to 0.5 millibars. If you have noted a rise or fall in air pressure during the previous three hours, mark a + or − accordingly.

Wind: Wind direction should be observed with the use of a wind vane or other indicator and recorded to eight points of the compass (N, NE, E, SE, S, SW, W, NW); wind speed should be estimated using the Beaufort scale (see pages 608–609) or an anemometer. If the Beaufort scale is used, enter the figure for speed after a hyphen following the direction (e.g., NE-2); if an anemometer is used, enter the miles or kilometers per hour after the direction (e.g., NW-33).

Rainfall: The amount of precipitation should be recorded at 12-hour intervals, with the figure to the nearest 0.01 inch (0.03 cm). If no precipitation has occurred, leave the column blank. If precipitation has occurred, but is less than 0.01 inch, enter "T" for "trace." Remember to empty your gauge after each observation.

Remarks: In the column labeled "Remarks," enter any special phenomena you have observed during the day, such as thunder and lightning, smoke, or haze, and indicate the times at which the events were observed. In addition, if there have been clouds, note here their approximate heights (low, medium, high), types, and directions of movement.

The next temperature-sensitive instrument that most amateur observers want is one that will automatically register the highest and lowest readings in a given period of time, usually 24 hours. Several types are available, including a longtime favorite known as the Six-type maximum-minimum thermometer, invented by Englishman James Six 200 years ago. It consists of a U-shaped tube with a creosote-filled bulb at one end. The creosote expands or contracts with temperature variation, pushing or pulling a short column of mercury in the lower part of the U-tube. Sliding iron indexes ride atop the twin mercury columns and remain at the extreme positions attained by each. The maximum scale, on the right, reads from low to high as in a regular thermometer; the minimum scale, on the left, is inverted. You can reset the indexes with the aid of a small magnet. Compute the mean temperature of any stated period (again, preferably once a day) by taking the sum of the maximum and minimum readings and dividing that number by two.

six-type maximum-minimum thermometer

Another type of maximum-minimum thermometer suitable for use by amateurs is called a digital thermometer. It employs an electric circuit whose resistance varies according to thermal change, and it gives readings to tenths of a degree. Of the many models available, the most common one displays outdoor and indoor temperatures simultaneously, in Fahrenheit or Celsius readings (controlled by the flip of a switch). A special memory feature preserves the maximum and minimum values registered since the last setting. A 10-foot (3-m) lead-in wire connects the unit with the temperature-sensitive bulb outside, which has a clip for securing it to an outdoor wall.

digital thermometer

Measuring Precipitation: A gauge to measure precipitation is the second type of instrument an amateur weather observer should acquire. The amount of local rainfall is of prime importance to farmers and gardeners, to city engineers and health officials, and, in fact, to anyone who uses water for bathing or drinking.

You can make a homemade rain gauge easily from a glass tumbler or a metal can with straight sides and a sharp-edged open rim at the top to split the raindrops. You need only leave this receptacle outside to collect rain and then measure the catch with a ruler to get a rough approximation of the amount of precipitation. For a more precise reading, select a receptacle whose opening measures ten times that of the measuring tube. When you pour the rainfall catch from the larger into the smaller vessel, the amount will be exaggerated ten times. Thus a measurement of 10 inches (25.4 cm) in the tube means that 1 inch (2.54 cm) was collected in the exposed receptacle. There are several rain gauges on the market that you will eventually want to possess. One such small-orifice gauge is made of clear plastic and can be purchased from a garden-supply or hardware store. Such gauges have a limited capacity, but they are inexpensive, convenient, easy to read, and fairly accurate, comparing favorably with larger ones, except in strong winds. They usually come with a support that can be used to fasten the gauge to a fence post or a stake set in the ground. However, these gauges are not suitable for measuring snow, since they clog with snowflakes and cannot be left exposed in below-freezing weather without suffering damage.

A more sophisticated clear-plastic rain gauge is also available. It is fashioned after the National Weather Service's standard 8-inch (20.3-cm) model and consists of a 4-inch-diameter (10.2-cm-

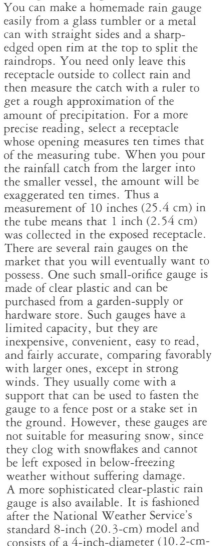

homemade rain gauge

plastic rain gauge

diameter) overflow collector with a fitted, knife-edged funnel of the same diameter. The funnel feeds the rainfall catch into a direct-read inner measuring tube that has an area one-tenth that of the funnel. The tube is imprinted with graduations to 0.01 inch (.03 cm) and has a capacity of 1 inch (2.54 cm). Excess rain is directed into the overflow collector, which will hold up to 10 inches (25.4 cm). For rainfalls greater than 1 inch (2.54 cm), you calculate total rainfall by filling the measuring tube with the water in the overflow collector as many times as necessary and totaling the amounts. The gauge comes with a stainless-steel bracket for easy attachment to a fence post or stake. This type of gauge can be used to measure snow as well. In below-freezing weather when snow is expected, remove the funnel and inner measuring tube and leave outdoors only the overflow collector. To measure the liquid equivalent of the snowfall, bring the snow-filled collector indoors, let its catch melt, and pour the water into the measuring tube. (Be aware that even gentle winds can create eddy currents around the outdoor gauge, diverting falling snow from the collector and causing an underestimate of the actual precipitation amount.)

long-term rain gauge

To directly measure the depth of snowfall in inches or centimeters, no instrument is needed except a yardstick or ruler. Take several measurements in an open yard or field where there is little apparent drifting and average them for a representative figure. Another method to calculate snowfall depth is to place a piece of plywood (about 6 feet/1.8 m or more square) on the grass of an open area and to measure the depth of each snowfall that accumulates on top of it. Picnic tables also work well as collecting surfaces. You can take a supplementary reading by using the 4-inch-diameter (10.2-cm-diameter) overflow collector. Turn it over and push it down on the deepest collection of snow on the board to extract a cross section of snow. Take the core indoors, melt it, and measure its water equivalent.

Measuring Atmospheric Pressure: The barometer, a time-honored aid of the weather-wise, has been greatly overrated as a forecasting device, but it does give you a good idea of your position on the weather map in relation to large-scale weather features. A barometer is a scale that weighs the column of air overhead. Since the density of the air determines the general circulation of the atmosphere, a rise or fall of the barometer indicates what type of air is moving in. When a southwest wind prevails aloft, it brings warm, moist air. Since warm, moist air weighs less than cold, dry air, the barometer falls, indicating an increased chance of rain or snow.

The basic barometer consists of a glass tube about 35 inches long. Its top is sealed and its open bottom is immersed in a cup of mercury. Tests have shown that the atmosphere pressing down on the surface of the mercury outside the tube will sustain the weight of a column of mercury inside the tube to a height, at sea level, that averages 29.92 inches. Barometer readings usually range from 29.00 to 31.00 inches.

The most useful barometer for an amateur is a small aneroid type (aneroid means "without fluid"). The heart of the instrument is a small metal cell in which there is almost no air. A strong spring within the cell prevents its walls from being crushed by atmospheric pressure. Small changes in air pressure tend to compress or extend the spring, whose slight movement is translated by linkages to a pointer that turns on a round face.

aneroid barometer

The "height" of the aneroid barometer is easily read from the dial face, which is marked in units representing inches or millibars. (Professional meteorologists now express atmospheric pressure in millibars, not inches of mercury. A bar is a unit of force or pressure that the atmosphere exerts. The average reading is 1013.25 millibars, and the usual range is from 980 to 1050 millibars.)

An aneroid barometer should not be placed outdoors because exposure to moisture and rain will cause corrosion. Since air pressure is the same inside your house as outside, place the instrument indoors at a convenient location. If you keep the barometer out of sunlight and away from drafts, in a place where temperature does not vary much, your readings will be reliable. To test the instrument's accuracy, keep a record of frequent readings and note any differences from radio reports in your immediate vicinity. You can make adjustments easily by turning the small screw at the back of the barometer. Also, your readings should be adjusted to sea level, just like those from the weather station.

An inexpensive substitute for an aneroid barometer is the water barometer, also known as the Cape Cod weather glass. It consists of a closed glass cistern, partially filled with water, with a swanlike, curved neck whose top is open to the air; it looks somewhat like a half-filled teapot. The water in the neck rises and falls with changes in atmospheric pressure. Because water expands and contracts with changes in temperature, you must keep a water barometer in a place with even temperature. The barometer's readings are not accurate or precise, but they do tell whether the air pressure is rising, falling, or remaining steady.

water barometer

Measuring Wind Direction and Speed: The wind, which is air in horizontal motion, is important for you to watch. Its direction can indicate what kind of weather is heading your way, and its speed can tell you how fast the changes are taking place.

The prevailing wind in most cases is a reflection of the regional weather pattern. You can determine wind direction by watching a wind vane, the flow of smoke, the behavior of a flag, or the trend of waves. A wind sock or a tethered Chinese kite with streamers will also tell you the direction of the wind. You can purchase an ornamental wind vane from a garden shop and mount it on your garage, or ask a next-door neighbor to mount it where it will be readily visible from your window. Should you choose to build your own small wind vane, remember to balance the tail vane with the direction arrow and to allow the vane to swing freely on a frictionless pivot.

wind sock

Observe the wind direction to eight points of the compass. Remember that wind direction refers to the direction the wind blows *from*. Thus, for example, a northeast (or northeasterly) wind blows *from* the northeast. If you do not have a compass, sight the North Star and mark its direction, or observe the shadow of the sun at noon, standard sun time, and orient your vane accordingly. The shifts in the wind are very significant, as any sailor will tell you, so keep track of the changes and the times at which they occur.

compass

Wind speed is a bit more difficult to gauge. You can get a good indication of relative speed by employing either a traditional Beaufort scale, which is based on visual signs and has been used extensively at sea since 1805, or its modification for land use (see pages 608–609). The original scale describes the actions of the wind on the waves, while the land version uses the angle of a smoke plume, the rustling of leaves, and the swaying of trees to specify various speeds. After a little practice watching the signs, you will be able to make good approximations of wind speed to enter in your weather log. Devices used to measure wind speed are called anemometers. One of the simplest types of anemometers is a wind sock mounted on a pole, a familiar sight at small airports and heliports. The wind sock tells you wind direction and gives a general idea of wind speed: The straighter the flying sock, the stronger the wind. You can make one from a basket ring and a piece of water-repellent cloth.

The real pride of a weather watcher is a mast mounted on the roof, fitted with a set of spinning cups that catch the wind and a vane that responds to every wind shift. Such anemometers and wind vanes usually register indoors on electrical dials.

Measuring Humidity:

sling psychrometer

Humidity (the amount of water vapor in the air) is an important property of the atmosphere. It not only determines what kind of weather we may have, it also regulates our relative comfort. The most common expression of humidity is relative humidity, the ratio of the amount of water vapor in the air to the amount the air would contain if it were fully saturated. For example, 100 percent relative humidity means the air is saturated and can hold no more moisture, while 50 percent means the air is holding only half its capacity. Any instrument that measures humidity is called a hygrometer; the standard hygrometer is a sling psychrometer, consisting of two matched thermometers—a "wet bulb" and a "dry bulb"—mounted on a steel or plastic backing. To provide a reading, the wet bulb is covered by a muslin sock that is moistened before being whirled in the air on a hand-held swivel or sling. The wet-bulb thermometer is cooled by the evaporation of the water on the sock. The drier the air, the greater the

The Beaufort Scale

Code	Wind Speed			Waves*
	(mph)	(kph)	(knots)	(height in feet)
0	<1	<1	<1	0
1	1–3	1–5	1–3	0
2	4–7	6–11	4–6	0–⅓
3	8–12	12–19	7–10	⅓–1⅔
4	13–18	20–29	11–16	2–4
5	19–24	30–38	17–21	4–8
6	25–31	39–50	22–27	8–13
7	32–38	51–61	28–33	13–20
8	39–46	62–74	34–40	13–20
9	47–54	75–86	41–47	13–20
10	55–63	87–101	48–55	20–30
11	64–74	102–120	56–63	30–45
12	75 +	120 +	64 +	45 +

Designation	Description
Calm	Smoke rises vertically; sea mirror-calm; tree leaves do not move
Light air	Smoke drift indicates wind direction; weather vane does not move
Light breeze	Wind felt on face; weather vane begins to move; leaves rustle; small wavelets
Gentle breeze	Wind extends light flags; leaves and twigs in constant motion; large wavelets
Moderate breeze	Dust and loose paper raised; small branches move; small waves; many whitecaps
Fresh breeze	Small trees sway; moderate waves
Strong breeze	Large branches move; wind whistles in wires; larger waves forming
Moderate gale	Whole trees move; walking affected
Fresh gale	Twigs break off trees; walking difficult; moderately high waves
Strong gale	Slight structural damage occurs; high waves; branches break
Whole gale	Trees uprooted; considerable structural damage; very high waves with overhanging crest
Storm	Widespread damage; extremely high waves; sea covered with white foam patches
Hurricane	Severe and extensive damage; visibility of sea greatly reduced

Height of waves are defined by the World Meteorological Organization

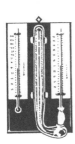

wet- and dry-bulb
hygrometer

evaporative cooling and the lower the wet-bulb reading. The dry bulb is unaffected by the whirling and simply measures current air temperature. Both thermometers are then read, and standard tables (usually supplied with a commercially obtained instrument) are consulted to determine relative and other humidity readings.

A related type of hygrometer employs wet- and dry-bulb thermometers mounted on a fixed wooden or plastic backing. Between them stands a water reservoir, which receives a wick from the wet bulb. The two bulbs are fanned vigorously with a piece of cardboard until the reading on the wet-bulb thermometer will descend no farther. The wet- and dry-bulb readings are then taken and tables are consulted. This type of psychrometer is easier to operate and less expensive than the sling psychrometer, but is not as accurate.

dial hygrometer

Even less accurate but easier to operate are dial hygrometers; these employ sensors made of materials, such as treated paper or human hair, that stretch, contract, or bend with changes in humidity. Like an aneroid barometer, the sensor is connected by various linkages to a dial pointer.

As mentioned earlier, relative humidity is not the only measurement of water in the air. Weather watchers also like to know the dew point—the temperature at which air is saturated and can hold no more moisture, and at which dew, fog, or clouds form. Like relative humidity, dew point is determined by taking wet- and dry-bulb readings and converting the data to dew point by consulting a set of standard tables (supplied with a commercial instrument).

Observing the
Present
Weather:

Present weather is a term employed by weather watchers to describe the current state of the sky and the presence or absence of precipitation or obstructions to visibility. There are, of course, countless varieties of present weather, each of which gives a clue to the type of air mass that controls the weather in the vicinity. A drizzle often indicates that a warm front is near and that warm air is overrunning a cold air mass. Shower activity usually points to overturning of tropical or polar air and instability—both characteristics of cold fronts. A change from one type of precipitation to another is significant. Changes from sleet to snow or from snow to rain, for example, give a hint of temperature changes in progress. As you make your observations, record them in your weather log.

Clouds: Clouds provide the best indications of present weather and its significance. The atmosphere is a great cloud factory, always at work, turning out countless varieties—sometimes beautiful formations to be admired and sometimes threatening masses that warn of dangers to come. It is in the realm of the clouds, whether as fog on the ground or as cirrus wisps blown off a cumulonimbus tower at heights of 50,000 feet (about 15 km), that an alert observer can get a definite knowledge of current developments. After a little study of the cloud photographs in this book or on a cloud chart and a few hours of practice observing actual clouds, you should be able to answer the following questions about clouds you see:
1. How much of the sky is cloud covered? (Use a scale from zero to ten.)
2. At what approximate height are the clouds—low (ground–6,500 feet/2,000 m), medium (6,500–16,500 feet/2,000-5,000 m), or high (above 16,500 feet/5,000 m)?
3. To what family or type do the clouds belong?
4. From what direction are the clouds

moving? Are different levels moving in different directions?

After you record your analysis of the clouds in the sky, follow the tips given in this essay in checking your thermometer, barometer, wind vane, anemometer, and humidity instruments for current readings and the changes that have taken place since your last observation. With this survey, you will be up-to-date on present weather and ready to attempt a weather forecast.

The Wind: You should always know from what direction the wind is coming because this tells whether the air will be cooler or warmer, drier or more moist, than the air you are experiencing.

In addition to the direction, the character of the wind should be noted. Is it shifting or holding steady? A gradual shift of the wind may be the first sign that an area of high pressure is about to move either north or south of you. That will make a big difference in the local weather during the next 24 hours. The approach of a front usually brings gradual shifts of wind; its passage may mean a quick, 90-degree-or-more shift in direction. Sometimes such a shift is the only evidence that a front has gone through your vicinity and that a new air mass is moving in. Watch the changes in wind speed rather closely. The wind predictably exhibits an increase in activity almost every afternoon, as the sun's heating takes effect, and a decrease in the evening about sunset. If the wind continues steadily through the afternoon and into the night, or if a steady wind is blowing when you rise in the morning, weather changes may soon be in store. Keep track of increases and decreases in wind speed over a period of several hours. An increase means that the barometric pressure is rapidly changing and that changes in the weather situation will be accelerated. On the other hand, a decrease in wind speed or no wind portends the continuation of the present type of weather.

Air Pressure: An aneroid barometer can be a useful guide, but you should put no faith in the traditional weather words that appear on the dial: Rain, Change, Fair. They are merely decorative and never appear on a professional instrument. Relying on these prophetic words will only teach the amateur forecaster some sad lessons. The importance of checking barometric pressure is not in looking at the actual reading on the dial, but in noting the barometric tendency, which is the amount of change that has taken place over the previous three hours. Ten classifications have been recognized, including *steadily falling, rapidly falling, falling then rising,* and *rising rapidly.* The barometric tendency, always reported with the other weather elements, is one of the first items you should consider in analyzing any report. Remember that it can rain when the barometer is at 30.50 inches, and it can be clear when the barometer is at 29.50 inches. The principal concerns of the forecaster are whether the air aloft is getting colder or warmer, and whether a surface low-pressure area is approaching. The barometer indicates much about these conditions.

A word of warning to new observers: There are barometric changes that occur periodically throughout each 24-hour period that have no relation to local conditions. Pressure waves move around the globe in much the same manner as daily tidal currents move in the ocean. At 10:00 A.M. and 10:00 P.M. local time, the barometer tends to peak; at 4:00 P.M. and 4:00 A.M., it reaches a low point. Although the changes are small—0.02 to 0.03 inch—they should be added or subtracted mentally in evaluating the reading of the barometer.

Wind and barometer observations work together in forecasting. After you have made your observations, refer to a wind-barometer table to get an idea of what future changes they may indicate. These indications apply in a general way to most of the country; your locality will have many differing local weather situations, but with a little experience, you should be able to begin your own wind-barometer chart, revising it until you have a fairly workable set of rules for your locale.

Temperature and Humidity: Always keep your eye on temperature and humidity changes since they give valuable clues to the type of air mass in your area. Temperature readings, of course, will vary greatly within 24 hours, as the sun's heat diminishes through radiation during the night. The effects of these daily fluctuations can be minimized by comparing temperature readings made at the same times each day. Day-to-day changes in these readings show more clearly changes in air mass than do readings made at different times of day. Above-normal humidities (high wet-bulb readings) in the morning indicate the presence of tropical air and the prospect of an oppressive day, perhaps with showers in the afternoon. The wet-bulb reading, you will find, is a better identifier of air-mass changes and frontal passages than temperature alone. If there is a big spread between dry-bulb and wet-bulb readings, it means the air is dry and the possibility of rain is quite remote; but if only two or three degrees separate the readings, watch out for fog or dew.

PROFESSIONAL WEATHER INSTRUMENTS

If you are really serious about weather observing, you may want to advance beyond the instruments discussed in the previous essay on becoming an amateur weather watcher to using some of the sophisticated instruments employed by professional forecasters. A great variety is available; each country has developed similar types of instruments, though they may vary in construction and appearance. Some instruments, such as Doppler radar and electronic psychrometers, are well beyond the means of the individual observer. However, many professional instruments are reasonably affordable. The following are descriptions of instruments employed by the National Weather Service of the United States and the Canadian Atmospheric Environment Service.

Temperature: The faulty air thermometer known as a thermoscope was first employed by Galileo circa 1593 in Florence. Alcohol-spirit thermometers were introduced in 1657 at the Academia del Cimento, also in Florence. Substantial improvements in thermometry were made in 1714 by Gabriel Fahrenheit, who perfected the mercury-in-glass thermometer and employed his own scale. The Celsius scale was established circa 1742 by Anders Celsius.

Official Thermometer: A seasoned glass tube, about 10½ inches (26.7 cm) long and usually filled with mercury, is attached to a stainless-steel backing. The tube has a long, cylindrical bulb at the bottom for maximum exposure to the surrounding air. The stem of the tube is etched with one-degree graduations in Fahrenheit or Celsius, with figures for each ten degrees also etched on the stem. The normal range of the thermometer is $-40°$ to $130°F$. For measuring temperatures below $-40°F$, alcohol is substituted for the mercury. A convenient brass support holds the thermometer and backing in proper position and can be attached to a crossboard inside the shelter.

Maximum-Minimum Thermometers: The maximum-minimum-thermometer set consists of a maximum-registering thermometer, a minimum-registering thermometer, and a metal support that holds the two and facilitates their resetting.
The maximum thermometer has a mercury-filled glass tube about 10 inches (25.4 cm) long and ¼ inch (1.91 cm) in diameter, with a bore of about 1/100 of an inch (0.0025 cm). A constriction in the tube, about an inch above the usually spherical bulb, acts much like a valve to permit the mercury to pass through in spurts as it expands with temperature increases. Upon a decrease in temperature, the mercury is unable to return through the narrow constriction. Thus, the column of mercury remaining in the upper part of the tube will indicate the highest temperature reached since the last resetting. The maximum thermometer can be reset by whirling it vigorously in a circle (the metal support supplies a fixed swivel for resetting). Centrifugal force drives the mercury back through

the constriction into the bulb. After resetting, the maximum thermometer should indicate the current temperature.

The tube of the minimum thermometer is of similar dimensions, but its bore is larger and filled with alcohol and with air under pressure. A floating index, consisting of a dumbbell-shaped piece of black glass about 9/16 of an inch (1.43 cm) long, rides in the column of alcohol. The index retreats with falling temperature because the surface tension at the end of the alcohol column draws it to the left. Upon an increase in temperature, the alcohol expands, destroys the surface tension, and flows around the index, leaving the right end of the index at the lowest reading. To reset the thermometer, tip it downward to the right so that gravity pulls the index back to the current temperature. The Townsend Support is an all-metal holder for the protection, proper exposure, and easy resetting of the two thermometers. Each thermometer is secured with clamps, enabling it to be moved without being dismounted.

Thermograph: A portable recording thermometer called a thermograph makes a permanent, inked record of temperature variations over a period of one to seven days, according to the time gears employed. The temperature-sensitive element is a Bourdon tube, an oblong tube of flexible metal filled with alcohol, whose expansion and contraction changes the curvature of the tube. (The Bourdon tube may be replaced with a bimetallic coil made of bonded strips of two different metals that expand and contract at different rates as the temperature changes.) One end of the curved tube (or bimetallic coil) is fixed, while the other end is free to move with temperature changes. The slight movement of the free end is transferred by a system of linkages to a pen arm, which traces a record on a chart. The chart is fastened to a vertical cylinder, rotated by a spring-driven clock mechanism, and measures 4⁵⁄₁₆ inches (10.5 cm) vertically for temperature and 11½ inches (29.2 cm) horizontally for time. The usual range of a standard thermograph is $-20°$ to $110°F$. The recording mechanism is enclosed in a metal case with a glass-paneled front, but the Bourdon tube is exposed to the open air on the side of the instrument.

Instrument Shelter:

The most commonly employed instrument shelter (described in "Becoming an Amateur Weather Watcher") is known as the cotton-region type. This medium-size shelter is the standard of the 12,000 cooperative observers who are the backbone of the National Weather Service's climatological network. It has four louvered sides, a ventilated floor, and a double roof with air space between the two layers. The interior measures about 23 by 21 by 18 inches (the Canadian standard shelter is slightly smaller) and is designed to house a set of maximum-minimum thermometers, a thermograph, and a sling psychrometer with whirler attachment, all of which are described in this section.

Humidity: The development of the familiar wet- and dry-bulb hygrometer to measure humidity is attributed to English geologist James Hutton in 1784. However, much of the theory of hygrometry and the invention of the hair hygrometer were announced by Horace Benedict Saussure of France in 1783.

Sling Psychrometer: A psychrometer is an instrument used to measure water vapor in the atmosphere. It consists of two thermally matched thermometers mounted on an aluminum backing that is fastened by a swivel attachment to a hardwood handle. The wet bulb, complete with a small muslin wick, free of sizing and starch, extends about 1½ inches below the dry bulb. After the wick is wet with clear water, the instrument is whirled vigorously in a circular motion, usually by hand, so that air passes over the two bulbs at a rate of at least 15 feet per second. When the mercury in the wet bulb will drop no farther, a reading of each thermometer is taken, and psychrometric tables are used to determine humidity or dew point. (See page 607 for details of this procedure.)

Hygro-thermograph: The hygrothermograph is an instrument that records both temperature and humidity on the same chart. The temperature-sensitive element is a Bourdon tube, as on a thermograph. Humidity measurements are regulated by several strands of human hair: An increase of moisture in the air causes the hair to expand, and a decrease causes it to contract. Through a system of linkages, the Bourdon tube and the hair cause separate recording pens to operate on a dual chart. Temperature is recorded in any desired range of 100 degrees (adjusted by turning a screw) on the upper two-thirds of the 5-inch vertical chart; relative humidity, from approximately 5 percent to 100 percent, is traced on the lowest third.

Atmospheric Pressure: The principle of the mercurial barometer was announced by Evangelista Torricelli, a pupil of Galileo, in 1643. The barometer soon became the standard instrument with which to determine atmospheric pressure. The aneroid (without fluid) method of measuring pressure was investigated by Nicholas Jacques Conte in 1795, and the first aneroid barometer was patented in 1845 by Lucien Vidie.

Aneroid The aneroid barometer is the
Barometer: meteorological version of the traditional
home barometer, which has the words
Fair, Change, and *Stormy* imprinted on
its face. The professional instrument
omits the weather words and usually
shows graduations in both inches of
mercury and in millibars. A small,
adjustable pointer enables you to judge
its movement since the last reading.
The pressure-sensitive element of an
aneroid barometer is a corrugated
chamber of thin nickel silver from
which most of the air has been
evacuated. One side of the chamber is
fastened to a strong spring, which holds
the vacuum chamber open. With a
decrease in pressure the spring opens,
and with an increase the spring closes.
These movements are transmitted and
magnified by a bar arm that moves a
delicate chain attached to the pointer
on the dial of the barometer.
Compensation for temperature changes
are made on all high-quality aneroids
by a special bimetal construction of the
bar arm.
A barometer indicates the pressure of
the air at its location (the station
pressure). For such a reading to be
compared with others on a weather
map, it must be converted to a sea-level
reading by applying a standard
correction for the altitude of the
instrument.

Microbarograph: The chief guide of the weather
forecaster is a recording barometer
called a microbarograph. Its cells,
sensitized to outside pressure by the
evacuation of most of the air from its
interior, create an inked record of
atmospheric pressure on an expanded
scale that can be read to 0.01 of an
inch. The operating element consists of
two aneroid cells, called sylphons,
suspended one over the other in a metal
cylinder. (Two cells are used to increase
sensitivity.) Each cell is constructed of

hard, silver-plated brass with a very thin but deeply grooved surface. A strong coil spring inside each cell prevents it from collapsing. The top of the upper cell is attached to an adjusting screw, while the bottom of the lower cell is free to move with changes in pressure. This movement is transmitted by a system of linkages and levers to a pen arm that records the pressure in inches or millibars. Charts measure 6¼ inches vertically for pressure and 11½ inches horizontally for time. The standard range is 28.50–31.00 inches or 960–1050 millibars, but range is adjustable. The recording period can be one, four, or seven days, according to the time gears employed.

Mercurial Barometer: The standard instrument for measuring atmospheric pressure in either the laboratory or the meteorological station consists of a glass tube about 35 inches long, with a sealed top and an open bottom. This glass tube, enclosed in a thin brass casing for protection, has a scale graduated in inches, millibars, or millimeters marked on the front. Inside the casing, a short sleeve serves as a vernier (a device to interpolate between graduations) from which fine readings to a thousandth of an inch can be obtained. On the outside of the casing, a small thermometer indicates the temperature in the vicinity of the mercury so that temperature corrections can be made. The bottom end of the barometer tube rests in a specially constructed leather cistern filled with mercury. The zero point of the barometer scale is indicated by an ivory pointer, which extends below the casing of the cistern. Before a reading is taken, the ivory pointer must be adjusted so it is just touching the level of the mercury in the cistern to establish a zero point on the barometric scale.

Wind Speed and Direction: In Athens in the second century A.D., a Tower of Winds was built, on which the eight directional winds were named and illustrated in sculpture. The first technical wind-speed indicator, devised by Robert Hooke of the Royal Society of London circa 1667, was of the pressure-plate type: a rectangular piece of metal that swings in the wind, its angle from the vertical giving an approximation of wind speed. Most often, however, wind speeds were simply estimated by observing the wind's effect on trees, flags, or water. Not until the invention of the rotating-cup anemometer in 1846 by T. R. Robinson, an Irish physicist and astronomer, were precise measurements of wind speed possible.

Anemometer:

The three-cup anemometer, regarded by the public as the symbol of the weather service, has long been the standard instrument of national meteorological services for measuring wind speed. The instrument consists of a tubular housing, a rotor assembly composed of three hemispherical cups mounted on radial spokes, and a gear system that transmits the movement of the rotor to electrical contacts and registering dials. The anemometer rotor and the gears are arranged in such a way that each $\frac{1}{10}$ of a mile of passing wind causes a complete revolution of the main gear. (One mile of wind equals 1 mph for 1 hour, 60 mph for 1 minute, etc.) There are six electrical contact points around a vertical gear, so one revolution causes six electrical contacts. Thus, one contact is closed for each $\frac{1}{60}$ of a mile of wind passing the station. Further, the number of contacts closing in one minute is equal to the speed of the wind in miles per hour: Sixty contacts per minute means 1 mile (1.6 km) of wind passing by each minute, or 60 miles per hour (97 kph).

The Aerovane System: Of the many different types of propeller wind-speed instruments available, the most popular in the United States consists of a bladed rotor and wind vane unit. The propeller, held directly into the wind by a streamlined vane resembling a wingless airplane, contains a small generator whose output is registered by a remotely located voltmeter (an instrument that measures electrical output in volts). The voltage is directly proportional to the speed of the rotor, which in turn is directly proportional to the speed of the wind. The rotor responds quickly to both gusts and lulls. The vane swivels on a shaft attached to a selsyn (self-synchronous) motor that transmits angular rotation to another selsyn motor in the indicator, or recorder. The Aerovane recorder employs a strip chart that can be set for different periods of recording: At normal speed the chart will trace a record for a period of two weeks without attention.

The Aerovane was originally designed for the U.S. Navy by the Bendix Aviation Corporation. The company is no longer in the meteorological-instrument business; however, the instrument is still made by several different companies.

Wind Vane: The standard weather-service wind vane consists of a carefully balanced aluminum rod 36 inches (91.4 cm) long with an arrow-shaped head and a modified fan-shaped tail. The vane is secured to a vertical shaft that rotates freely within housing that contains a ring set securely in the main shaft. Around the circumference of the ring are either four or eight electrical contact points matching four or eight compass directions. As the vane and shaft rotate, one of the contacts closes and permits an electrical current to pass. This lights one of the four or eight lamps on a remote indicator, informing the observer instantaneously of wind direction. If two adjacent contact points are touched simultaneously, two lamps will light, allowing you to read wind direction to 8 or 16 points of the compass.

Wind Indicator:

The wind indicator consists of a circle of eight lamps and one central lamp, enclosed in a round case with a metal frame and dial. Wind direction is indicated by the eight lamps, as described above. The center lamp and a buzzer are wired to the wind-speed transmitter to flash and sound once for each $\frac{1}{60}$ of a mile of wind rotating the cups. The number of flashes or buzzes per minute equals the wind speed in miles per hour.

Precipitation:

Rain gauges are known to have been employed in the ancient Korean Empire at least as early as A.D. 1442. The first recording rain gauge was introduced in 1662 by Sir Christopher Wren, the noted English architect and member of the Royal Society of London. Regular records of rainfall were made in England as early as 1667 and have continued to be taken to the present.

Rain-and-Snow Gauge:

The standard instrument of U.S. weather services for measuring rainfall and snowfall has not changed since 1870. Known as a stick gauge, it has four main components: a collector ring and funnel, an inner measuring tube, an overflow can, and a measuring stick. The collector ring is made of seamless brass tubing with an inside diameter of eight inches. The outside of the rim is beveled to a sharp, knifelike edge (to split raindrops falling on the rim instead of wholly in the collection area). The funnel is made of copper and tapers to a $\frac{5}{8}$-inch opening at the bottom to prevent excessive evaporation of rainfall from the measuring tube after a storm. The overflow can, also made of copper, has an inside diameter of 8 inches and a height of 24 inches; it serves to contain rainwater that overflows from the measuring tube in heavy downpours. The measuring tube, made of seamless steel tubing, has an inside diameter of 2.53 inches and is 20 inches high. The areal cross section of this tube is one-tenth that of the collector ring. Consequently, the rainfall collected in the tube will stand ten times as deep as what actually fell, making accurate measurements of even small amounts possible. A red cedar or plastic measuring stick, 24 inches long and marked to show every $\frac{1}{100}$ of an inch of rain, is the final component of the gauge.

Recording Rain-and-Snow Gauge: Another instrument to measure the amount of rain or snow employs the principle of weight. It has a knife-edge collector rim with an inside diameter of 8 inches, constructed of nonferrous material. The water catch is directed to a 12-inch capacity bucket that is mounted on a weighing mechanism (similar to a spring-loaded scale). This mechanism converts the weight of the precipitation to its equivalent in inches of water and activates a pen arm to trace a record on the chart. The gauge will make records covering 6, 12, 24, 48, 96, or 192 hours, depending on the type of time gears and charts employed. In below-freezing weather, special chemicals, usually automobile antifreeze, are placed in the bucket to melt snow and ice so that the water content may be measured.

Tipping-Bucket Rain Gauge: This instrument, consisting of a collector ring with funnel, an overflow reservoir, and a tipping-bucket mechanism, all mounted on a tripod support, is designed to report each $\frac{1}{100}$ of an inch of rain that falls. The collector has a rim of seamless tubing, 12 inches in diameter, beveled to a sharp edge at the top. The funnel portion of the collector narrows to a

small outlet immediately above the bucket mechanism. The tipping bucket is divided into two equal compartments, each holding exactly $\frac{1}{100}$ of an inch of rain. When one compartment fills, the bucket tips, momentarily closing a mercury-in-glass electrical contact, and empties into the overflow reservoir. The opposite compartment of the bucket mechanism is simultaneously positioned below the nozzle to continue receiving incoming rainfall. Electrical impulses are transmitted to a recorder, where each $\frac{1}{100}$ of an inch is recorded on a time chart.

GLOSSARY

Below are some meteorological terms and
North American weather features.

Acid precipitation	Rain or snow with a pH value lower than 5.6.
Adiabatic temperature change	The cooling or warming of air caused by expansion or contraction. The rate of temperature change in saturated air (the wet adiabatic rate) is always less than the rate of change in dry air (the dry adiabatic rate).
Advection	Horizontal movement of any atmospheric element: air, moisture, or heat.
Air mass	An extensive body of air whose horizontal distribution of temperature and moisture is nearly uniform.
Airstream	A substantial body of air with the same properties flowing along with the general circulation.
Alaskan Winds	Winds blowing down Alaskan valleys; they have local names, such as Knik, Matanuska, Pruga, Stikine, Taku, Take, Turnagain, and Williwaw.
Alberta clipper	A small, fast-moving cyclonic storm that forms on the Pacific front, usually over the Rocky Mountains of Alberta, Canada, and travels southeast into the Great Plains; it is followed by an outbreak of cold polar air.

Aleutian Low A low-pressure center, usually located south of or over the Aleutian Islands of Alaska, that forms one of the major centers of action in the air circulation of the Northern Hemisphere. Here temperate and polar winds meet and low-pressure systems generate or rejuvenate.

Anticyclone An area of high atmospheric pressure characterized by a center of maximum pressure and outflowing winds blowing clockwise (in the Northern Hemisphere). Also called a high-pressure system or a high.

Aphelion The point in the orbit of a planet that is farthest from the sun.

Arctic air An air mass that has been conditioned over high latitudes. Arctic air is colder and drier than polar air.

Arctic front The leading edge of arctic air in northern latitudes where it meets polar air or modified tropical air from the south.

Arctic sea smoke Steam fog rising from a body of (unfrozen) water as a result of very cold air passing over it.

Atmospheric pressure The weight per unit of the total mass of air above a given point. Also called barometric pressure.

Azores-Bermuda High A semipermanent high-pressure area in the north Atlantic Ocean that migrates east and west with varying central pressure. In summertime it causes a southerly circulation over the Atlantic states, bringing heat and humidity northward.

Backing wind A wind that shifts counterclockwise, such as from northeast to north to northwest.

Bane of Boulder	Winds descending the Front Range of the Rocky Mountains in the vicinity of Boulder, Colorado, caused by high pressure building over the Intermountain region and the western Rockies. Most frequent and strongest in winter.
Barometric pressure	See Atmospheric pressure.
Barometric tendency	The amount and character of a change in the barometer readings over any three-hour period.
Barrier Winds	A westerly flow of air along the northern slope of the Brooks Range in northern Alaska, preceding the arrival of cold air from the north.
Bellot	Winds blowing through the narrow Strait of Bellot, connecting the Gulf of Boothia and Franklin Strait in the Canadian high arctic.
Black blizzard	A severe dust storm on the Great Plains that darkens the sky and casts a pall over the land. Often accompanied by a sharp cold front. Also called a black roller.
Blizzard	A severe winter storm characterized by low temperatures (10°F/−12.2°C or below), wind speeds of 32 mph (51 kph) or higher, blowing snow, and visibility of 500 feet (152 m) or less.
Blue norther	An outbreak of cold air in Texas marked by a dark, blue-black sky.
Boulder wind	An especially strong downslope wind in the Front Range of the Rockies in the area of Boulder, Colorado.
Canyon wind	A flow of air caused by varying thermal conditions in different parts of a canyon. In daytime, heating near the mouth of the canyon causes an upslope

flow of warm air; at nighttime, cooling at the head of the canyon causes a downslope flow of cool air.

Catalina eddy   A weak cyclonic circulation formed off Point Conception in Southern California that turns toward land in the vicinity of Santa Catalina Island.

Ceiling   The height of the lowest layer of clouds or obscuring phenomena when $\%_{10}$ or more of the sky is covered. When less than $\%_{10}$ covered, the ceiling is termed unlimited.

Chinook   A warm downslope wind, or foehn, especially on the eastern slope of the Rocky Mountains. It generally blows from the southwest and is accompanied by a marked rise in temperature. Also called snow eater because in wintertime it can melt the snow over which it passes and leave the ground bare.

Chocolatero (also called a chocolate gale)   A moderate northerly wind in the Gulf of Mexico off Mexico; related to the Texas blue norther.

Chocolatta north   A northwesterly gale of the West Indies.

Circulation   The flow pattern of moving air. The general circulation is the large-scale flow characteristic of the semipermanent pressure systems in the atmosphere, while the secondary circulation occurs in more temporary, migrant high- and low-pressure systems.

Circulation cells   Large areas of air movement confined to a specific region that control the type of weather prevailing there.

Cloudburst   In popular terminology, any sudden and heavy fall of rain, almost always of the shower type. Also called a rain gust or rain gush.

Coastal storm A low-pressure disturbance forming along the South Atlantic coast and moving northeast along the Middle Atlantic and New England coasts to the Atlantic Provinces of Canada. Also called a northeaster.

Cold front The leading edge of an advancing cold air mass that is displacing a retreating warmer air mass.

Collada A strong, steady norther in the Gulf of California.

Colorado Low A low-pressure disturbance that forms in the lee of the Rocky Mountains, usually in southeast Colorado.

Condensation The process whereby a substance changes from vapor to liquid or solid.

Condensation nuclei Microscopic atmospheric particles on which water vapor condenses.

Conduction Transmission of heat by direct contact.

Convection In meteorology, the circulating motion of warm and cold currents of air; predominantly seen in the vertical motion of warmer air.

Convergence A distribution of wind movement that results in a net inflow of air into a particular region.

Coriolis force The deflective effect of the earth's rotation on all free-moving objects, including the atmosphere and oceans. Deflection is to the right in the Northern Hemisphere and to the left in the Southern Hemisphere.

Coromell A nocturnal land breeze prevailing between November and May in the vicinity of La Paz, near the lower end of Baja California, Mexico.

Cow-killer Strong east to northeast winds descending from the mountains in eastern Washington. Also called a Palouser.

Crown-of-winter storm   A snowstorm in early March that raises the cover of snow on the ground to the maximum depth of the winter.

Cyclogenesis   The process leading to the development of a new low-pressure system or to the intensification of a preexisting one.

Cyclone   A system of atmospheric pressure characterized by minimum pressure at its center and winds blowing inward (counterclockwise, in the Northern Hemisphere). Also known as a low-pressure system or a low.

Cyclonic flow   In the Northern Hemisphere, winds blowing counterclockwise.

Daily mean   A temperature reading obtained by averaging the maximum and minimum temperatures of a 24-hour period from midnight to midnight, or sometimes by averaging the hourly readings.

Denver cyclone   A small cyclonic circulation that develops a short distance north and east of Denver. It occurs when south to southwest winds curl around a low ridge south of the city. Its winds are only about 10 miles per hour, but the rotational motion encourages the growth of thunderstorms in the vicinity.

Deposition   The process whereby water vapor changes directly to ice without going through the liquid state.

Depression   An area of low atmospheric pressure.

Dew   Water droplets that form on objects as a result of radiational cooling and condensation of water vapor from the air.

Dew point The temperature at which a parcel of air reaches moisture saturation as it is cooled at a constant pressure.

Discontinuity In meteorology, the rapid variation of the gradient of an element, such as the reading of pressure or level of temperature at a front.

Disturbance An area of low pressure attended by storm conditions.

Divergence A distribution of wind movement that results in a net outflow of air from an area.

Dog days A period of especially hot summer weather extending from July 3 to August 10, when the Dog Star, Sirius, lies in conjunction with the sun; the star was once believed to add its intensity to that of the sun, creating high temperatures.

Doppler radar A type of radar used in meteorology to detect the motion of an air parcel directly.

Downburst A concentrated, strong downdraft from a severe thunderstorm that induces an outward burst of often damaging winds at the surface.

Downdraft A sudden descent of a stream of cool air from aloft, often causing windshear.

Drought A prolonged and damaging condition of subnormal precipitation in a given area.

Dry northeaster A steady wind from the northeast on the South Atlantic coast that originates in a high-pressure system centered over or near New England. It causes a cool period with high seas, especially along the north Florida coast.

Dust Bowl A term applied to the area of the Great Plains that was affected by the Great

Drought of the 1930s. The center was in the contiguous parts of Texas, Oklahoma, Kansas, Colorado, and New Mexico.

Dust devil   A small whirling column of wind caused by unequal heating of the earth's surface. It picks up loose dirt and leaves as it prances across the countryside.

Dust storm   A local or extensive storm with strong winds that raise small particles of soil into the air; often severe in droughty regions. Sometimes called by the Sudanese name *haboob*.

Easter   An easterly flow of dry winds across western Oregon.

Equinoctial storm   A disturbance at the time of the vernal or autumnal equinox, once thought to result from the sun's crossing the equator.

Evaporation   The change in a substance from liquid or solid to a gaseous state.

Extratropical cyclone   A low-pressure disturbance in middle or northern latitudes with a system of fronts.

Eye   In meteorology, the roughly circular area of lowest pressure, relatively light winds, and often fair weather at the center of a tropical storm.

Eye wall   The vertical bank of clouds, containing high winds, that surrounds the eye of a tropical storm.

Filling   The increase in pressure at the center of a storm system.

Flash flood   A sudden rise in the level of water in a small river or stream, brought on by intense rainfall.

Foehn   See Chinook.

Fog A visible aggregate of minute water droplets suspended in the atmosphere near the surface of the earth.

Freezing The transformation of a substance from a liquid to a solid state.

Freezing rain Rain falling through above-freezing air that freezes upon impact with a surface surrounded by below-freezing air.

Front The interface or transition zone between air masses of different properties.

Frontogenesis The formation and development of a front.

Frontolysis The deterioration and disappearance of a front.

Frost Ice crystals formed on grass or other objects by the sublimation (direct transfer) of water vapor from the air at below-freezing temperatures.

Glacier winds Downslope, sometimes high-speed winds descending from Alaskan glaciers.

Glaze A coating of ice on objects, resulting from an ice storm.

Greenhouse effect The heating of the lower atmosphere that results when shortwave solar radiation reflected from the surface of the earth is trapped in the atmosphere by water vapor and carbon dioxide.

Gully-washer A sudden heavy shower.

Haboob See Dust storm.

Hail Precipitation produced by cumulonimbus clouds in the form of balls or lumps of ice.

Haze Fine dust, salt, pollen, or other minute particles dispersed through the atmosphere that reduce visibility and

give the atmosphere a whitish
appearance.

High-pressure   See Anticyclone.
system

Hudson Bay   A high-pressure area situated over or
High   near Hudson Bay, that can profoundly
affect weather patterns by introducing
very cold air to the Northeast or
blocking storm movements.

Hurricane   A severe tropical disturbance in the
north Atlantic Ocean, Caribbean Sea,
or Gulf of Mexico that achieves a
sustained wind force of at least 74 mph
(118 kph).

Icelandic Low   A center of cyclonic action (usually
southwest of Iceland) where polar and
temperate airstreams meet and low-
pressure systems are generated or
rejuvenated.

Indian summer   A period of warmer-than-normal
temperatures, cool nights, sunny but
hazy skies, and lowered visibilities that
occurs after the first killing frost of
autumn.

Instability   A condition of the atmosphere whereby
a parcel of air given an initial vertical
impulse will tend to continue to move
upward. The upward current often
forms cumulus clouds that may grow
into thunderstorms.

Intermountain   A late fall or winter high-pressure area
High   between the Rocky Mountains and the
Sierra-Cascade range; it often blocks the
eastward movement of Pacific cyclones.
Also called Plateau High or Great Basin
High.

Inversion   A reversal of the normal arrangement
whereby temperature decreases with
increasing altitude, resulting in warmer
air overlying colder air; also, the layer
in which this reversal occurs.

Ionosphere   A complex atmospheric zone of ionized gases between altitudes of 48 and 240 miles (77–386 km).

Isobar   A line drawn on a weather map connecting points of equal barometric pressure (as adjusted for elevation).

Isohyet   A line drawn on a weather map connecting places with equal rainfall over a specific period of time.

Isotherm   A line drawn on a weather map connecting points of equal temperatures.

Jet stream   A zone of relatively strong winds concentrated in a narrow zone of the upper atmosphere, usually between 25,000 and 45,000 feet (8,000 and 14,000 m) altitude.

Lake-effect mechanism   A turbulent action of the atmosphere when a cool airstream passes over a warm body of water; it may result in rainshowers or snow showers, sometimes heavy, on the lee shore of a lake.

Lifting   The forced ascent of an airflow by a rise in terrain or by an air mass of denser properties.

Low-pressure system   See Cyclone.

Marine layer   A zone of the lower atmosphere that passes over an ocean or large body of water.

Mean temperature   The average of any series of temperature readings taken over a period of time. See also Daily mean.

Mesosphere   The zone of the atmosphere between the stratopause and the mesopause.

Mono winds   An airflow along the eastern flank of the Sierra Nevada near Lake Mono.

Monsoon   The flow of air from ocean to continent, mainly in summer.

Newhall winds   Downslope winds blowing from desert uplands through Newhall Pass into the San Fernando Valley north of Los Angeles.

Normal   The average value of a meteorological element over a fixed period of years.

Northeaster, or nor'easter   See Coastal storm.

Norther   A strong north wind on the Great Plains that follows the passage of a cold front.

Northwester, or nor'wester   A strong northwesterly wind, usually following the passage of a sharp cold front and the approach of a strong high-pressure area.

Occluded front   A composite of two fronts produced when a cold front overtakes a warm front and forces the warm air aloft.

Orographic lifting   A flow of air as it is forced to ascend to a higher elevation by rising terrain.

Overrunning   The ascent of warm air over cooler air, usually in advance of a warm front.

Pacific High   A large high-pressure system extending along the middle latitudes of the Pacific Ocean north of the Hawaiian Islands.

Palousers   Dangerous winds in the Palouse River valley in northern Idaho and eastern Washington.

Perihelion   The point in the orbit of a planet that is closest to the sun.

Plow wind   Strong, straight-line winds in the American Midwest associated with squall lines and thunderstorms. Also known as plough wind.

Polar air An air mass conditioned over tundra or snow-covered terrain of high latitudes.

Polar front A semipermanent discontinuity separating cold, polar easterly winds and relatively warm westerly winds of the middle latitudes.

Pressure gradient The change in atmospheric pressure over a given distance at a particular time.

Radiation The transfer of heat (or other forms of energy) through space without the agency of intervening matter.

Refraction The bending of light as it passes obliquely from one transparent medium (such as air) to another of different density (such as crystals).

Relative humidity The ratio (expressed as a percentage) of the actual amount of water vapor in the air to the amount the air could hold if saturated and if the temperature remained constant.

Ridge An elongated area of high atmospheric pressure.

Santa Ana A wind originating over the elevated deserts of eastern Southern California, warming rapidly in its descent through mountain passes to the coastal plain, and arriving as a hot, very dry, gusty wind.

Saturation The condition in which a parcel of air holds the maximum possible amount of water vapor.

Sea turn A New England term for an east wind from the cool offshore waters; usually attended by low clouds, fog, and sometimes drizzle. Occurs most frequently in springtime and early summer.

Secondary cold front A cold front that sometimes follows a primary front and brings an even colder air mass.

Secondary depression An area of low pressure that forms in a trough to the south or east of a primary storm center.

Semipermanent high or low A relatively stationary and stable pressure-and-wind system. Examples are the Icelandic Low and the Bermuda High.

Siberian express A popular term for fierce, cold cyclonic storms that descend from Alaska and northern Canada into the United States.

Sleet Small pellets of ice formed when raindrops fall through a layer of below-freezing air and congeal. Sleet often falls in a transition zone between rain and snow.

Snow Precipitation in the form of white or translucent ice crystals, chiefly in complex branched hexagonal form and clustered into snowflakes.

Sonora A term applied to a wind flow originating in Sonora, Mexico, and moving to Southern California and Baja California.

Source region An area of nearly uniform surface characteristics over which a large body of air stagnates and acquires a more or less equal horizontal distribution of temperature and moisture.

Southeaster, or sou'easter A strong southeasterly wind or gale that precedes a low-pressure disturbance.

Southwester, or sou'wester A strong southwesterly wind preceding the passage of a cold front.

Squall line A well-marked line of instability, usually accompanied by gusty winds, turbulence, and often heavy showers.

Stationary front A zone where cold and warm air meet that exhibits little or no movement.

Storm track The path usually followed by cyclonic disturbance in a region.

Stratopause The boundary zone between the stratosphere and the mesosphere.

Stratosphere The zone of the atmosphere above the troposphere and below the mesosphere, characterized at lower altitudes by isothermal conditions and at higher ones by gradually increasing temperatures. Ozone is concentrated here.

Sublimation The process whereby a solid changes directly to a gas, or sometimes vice-versa, without passing through the liquid state.

Subsidence The descent of a body of air, usually within a high-pressure area, that causes a spreading-out and warming of the lower layers of the atmosphere.

Supercooled The condition of water droplets that remain in a liquid state at temperatures well below freezing.

Temperature-humidity index (THI) A guide to human comfort or discomfort based on conditions of temperature and relative humidity.

Tendency The local rate of change of a meteorological element; see also Barometric tendency.

Texas Low A low-pressure disturbance that originates over Texas and pursues a northeasterly course. A Texas Low can create strong thermal contrasts along its tracks and cause severe local weather.

Texas norther A strong flow of north or northwest winds following a sharp cold front on the southern Plains. Temperatures may drop 20° to 30°F in a few minutes in wintertime. See Blue norther.

Tornado   A violently rotating column of air, pendant from a cumulonimbus cloud and nearly always visible as a "funnel cloud." Smaller tornadoes are popularly called twisters. On a local scale, a tornado is the most destructive of atmospheric phenomena. Its vortex is commonly several hundred yards in diameter, and it whirls, usually cyclonically, with wind speeds estimated from 100 to 300 mph (160–480 kph). Tornadoes occur most frequently in the late afternoon and evening hours from April to June, and they are accompanied by thunderstorms. Their path is usually from southwest to northeast. About 745 tornadoes are reported each year in the United States.

Trace   Describing an amount of precipitation less than 0.005 inch (0.127 mm).

Tropical air   An air mass conditioned over land or ocean within the tropics. It is usually warm and moist. When tropical air is carried north to clash with polar air, storms generate in the meeting zone.

Tropical depression   A cyclonic disturbance that originates over tropical waters and whose wind speeds do not exceed 38 mph (61 kph).

Tropical storm   A cyclonic disturbance that originates over tropical waters and whose maximum wind speed does not exceed 74 mph (118 kph).

Tropopause   The boundary zone between the troposphere and the stratosphere.

Troposphere   The lowest layer of the atmosphere.

Trough   An elongated area of low pressure.

Unstable air Air that does not resist vertical displacement. If unstable air is lifted, its temperature will not cool as rapidly as that of the surrounding air, and it will continue to rise on its own. An example is the rising column of air under a cumulus cloud.

Updraft An upward movement of air, usually in a thunderstorm, resulting from heating of a cumulonimbus cloud and greatly increasing in-cloud turbulence.

Vapor pressure That part of the total atmospheric pressure attributable to water vapor content.

Veering wind A wind that shifts clockwise—that is, from southwest to west to northwest.

Virga Streaks of water or ice particles that fall from a cloud and evaporate before reaching the ground.

Visibility The horizontal transparency of the atmosphere.

Vorticity The tendency of air (or other substances) in motion to rotate.

Warm front A leading edge of advancing warm air that displaces cooler air at the earth's surface or overruns it.

Warm sector The portion of a cyclone containing warm air; usually located in the eastern or southeastern section of a storm system.

Wasatch winds Strong, easterly, jet-effect winds blowing out of the canyon mouths of the Wasatch Mountains in Utah. These winds sometimes reach hurricane force on the plain and cause property damage.

Washoe zephyr A westerly chinook descending the eastern flank of the Sierra Nevada Mountains.

Waterspout   A tornado-like formation over water, usually much smaller and less vigorous than a true tornado. Waterspouts tend to occur most frequently in tropical waters, although they have been observed on the Great Lakes and on other northern lakes in Canada and the United States.

Water vapor   Atmospheric moisture existing as a gas.

Wave   A usually small cyclonic center in the early stage of development that moves along a cold front. Also referred to as a wave cyclone.

West Virginia High   A stagnant high-pressure area over West Virginia during Indian summer in the East.

Whirlwind   A general term for a small-scale, rotating column of air.

Whiteout   A condition occurring in the Arctic regions on a sunless day when clouds and surface snow seem to merge so that no horizon is perceptible. The term is also applied to blizzard conditions in which blowing snow restricts visibility to near zero.

Wind   Air in motion parallel to the surface of the ground.

Wind-chill factor   A figure used to express the cooling effect of the combination of particular temperatures and air speeds on exposed human skin.

Windshear   A sudden variation in the vector of wind flow that is especially dangerous to aircraft during takeoff and landing.

PICTURE CREDITS

The numbers in parentheses are plate numbers. Some photographers have pictures under agency names, which appear in boldface. Photographers hold copyrights to their works.

Steve Albers (168, 169, 218, 219, 222, 230)
American Red Cross (369, 372)
Annie Griffiths Belt (174, 237)
David O. Blanchard (208, 228)
Gary Braasch (137, 142, 177, 178, 180, 263)
Jay Brausch (133, 361, 363)
Ed Cooper (15, 96, 100, 141, 282, 283, 310)
Betty Crowell, (24, 281, 313, 314)
Mary A. Davis/Yellowstone National Park (307)

Marvin L. Dembinsky, Jr. Photography Associates
Marvin L. Dembinsky, Jr. (134, 318)
Skip Moody (172, 175)
Stan Osolinski (19, 37, 93, 260, 305, 350, 352, 354)
Ken Scott (124)
Joe Sroka (83)

Peter Dodge, National Hurricane Center/Hurricane Research Division (248)

Earth Images
C. Johnny Autery (193)
Chris Chandler (258)
Greg Edmonds (257)
David Gardiner (88)
Joel W. Rogers (285)

Arjen & Jerrine Verkaik/SKYART (21,
22, 26, 28, 29, 30, 32, 35, 42, 47,
49, 50, 51, 58, 59, 66, 74, 75, 78,
80, 81, 84, 85, 86, 87, 91, 95, 99,
102, 104, 105, 106, 107, 108, 109,
110, 120, 122, 125, 126, 127, 128,
130, 132, 158, 159, 160, 161, 162,
163, 164, 181, 195, 200, 215, 216,
223, 224, 234, 272, 293, 297, 335,
345)
Voscar/The Maine Photographer (334)
Tom Walker (337)
Steve Warble (16, 62, 65, 67, 71, 77,
89, 143, 167, 183, 185, 271, 288,
311)
John W. Warden (328, 359, 360, 364)
Washington State Historical Society
(368)
Fred Whitehead (36, 63, 319)

Wide World Photos
Gary Dunkin (378)
Charles L. Franck (373)

Steven C. Wilson/ENTHEOS (145)
Kent Wood (39, 45, 101, 139, 150,
190, 233)
Yellowstone National Park (303, 306)

INDEX
Numbers in boldface type refer to color plates. Numbers in italics refer to pages.

NATIONAL AUDUBON SOCIETY
FIELD GUIDE SERIES

Also available in this unique all-color,
all-photographic format:

African Wildlife • **Birds** *(Eastern Region)* • **Birds**
(Western Region) • **Butterflies** • **Fishes, Whales,
and Dolphins** • **Fossils** • **Insects and Spiders** •
Mammals • **Mushrooms** • **Night Sky** • **Reptiles
and Amphibians** • **Rocks and Minerals** • **Seashells**
• **Seashore Creatures** • **Trees** *(Eastern Region)* •
Trees *(Western Region)* • **Tropical Marine Fishes**
• **Wildflowers** *(Eastern Region)* • **Wildflowers**
(Western Region)

Prepared and produced by Chanticleer Press, Inc.

Founding Publisher: Paul Steiner
Publisher: Andrew Stewart

Staff for this book:

Senior Editor: Ann Whitman
Managing Editor: Barbara Sturman
Editors: Jane Mintzer Hoffman, Carol Healy
Editorial Assistants: Michaela Porta, Kate Jacobs
Project Assistant: Ferris Cook
Production: Gretchen Bailey Wohlgemuth
Designer: Barbara Balch
Photo Editor: Timothy Allan
Drawings: Edward Lam
Thumb Tab Symbols: Robert Fontaine
Layout Assistant: Sheila Ross
Original series design by Massimo Vignelli

All editorial inquiries should be addressed to:
Chanticleer Press
665 Broadway, Suite 1001
New York, NY 10012

To purchase this book, or other National Audubon Society
illustrated nature books, please contact:
Alfred A. Knopf
201 East 50th Street
New York, NY 10022
(800) 733-3000